McGRAW-HILL NETWORKING AND TELECOMMUNICATIONS

Build Your Own

Trulove — *Build Your Own Wireless LAN* (with projects)

Crash Course

Louis — *Broadband Crash Course*
Vacca — *I-Mode Crash Course*
Louis — *M-Commerce Crash Course*
Shepard — *Telecom Convergence, 2/e*
Shepard — *Telecom Crash Course*
Bedell — *Wireless Crash Course*
Kikta/Fisher/Courtney — *Wireless Internet Crash Course*

Demystified

Harte/Levine/Kikta — *3G Wireless Demystified*
LaRocca — *802.11 Demystified*
Muller — *Bluetooth Demystified*
Evans — *CEBus Demystified*
Bayer — *Computer Telephony Demystified*
Hershey — *Cryptography Demystified*
Taylor — *DVD Demystified*
Bates — *GPRS Demystified*
Symes — *MPEG-4 Demystified*
Camarillo — *SIP Demystified*
Shepard — *SONET / SDH Demystified*
Topic — *Streaming Media Demystified*
Symes — *Video Compression Demystified*
Shepard — *Videoconferencing Demystified*
Bhola — *Wireless LANs Demystified*

Developer Guides

Vacca — *I-Mode Crash Course*
Guthery — *Mobile Application Development with SMS*
Richard — *Service and Device Discovery: Protocols and Programming*

Professional Telecom

Smith/Collins — *3G Wireless Networks*
Bates — *Broadband Telecom Handbook, 2/e*
Collins — *Carrier Grade Voice over IP*
Harte — *Delivering xDSL*
Held — *Deploying Optical Networking Components*
Minoli/Johnson/Minoli — *Ethernet-Based Metro Area Networks*
Benner — *Fibre Channel for SANs*

Bates	*GPRS*
Sulkin	*Implementing the IP-PBX*
Lee	*Lee's Essentials of Wireless*
Bates	*Optical Switching and Networking Handbook*
Wetteroth	*OSI Reference Model for Telecommunications*
Russell	*Signaling System #7, 4/e*
Minoli/Johnson/Minoli	*SONET-Based Metro Area Networks*
Nagar	*Telecom Service Rollouts*
Louis	*Telecommunications Internetworking*
Russell	*Telecommunications Protocols, 2/e*
Minoli	*Voice over MPLS*
Karim/Sarraf	*W-CDMA and cdma2000 for 3G Mobile Networks*
Bates	*Wireless Broadband Handbook*
Faigen	*Wireless Data for the Enterprise*

Reference

Muller	*Desktop Encyclopedia of Telecommunications, 3/e*
Botto	*Encyclopedia of Wireless Telecommunications*
Clayton	*McGraw-Hill Illustrated Telecom Dictionary, 3/e*
Radcom	*Telecom Protocol Finder*
Pecar	*Telecommunications Factbook, 2/e*
Russell	*Telecommunications Pocket Reference*
Kobb	*Wireless Spectrum Finder*
Smith	*Wireless Telecom FAQs*

Security

| Nichols | *Wireless Security* |

Telecom Engineering

Smith/Gervelis	*Cellular System Design and Optimization*
Rohde/Whitaker	*Communications Receivers, 3/e*
Sayre	*Complete Wireless Design*
OSA	*Fiber Optics Handbook*
Lee	*Mobile Cellular Telecommunications, 2/e*
Bates	*Optimizing Voice in ATM / IP Mobile Networks*
Roddy	*Satellite Communications, 3/e*
Simon	*Spread Spectrum Communications Handbook*
Snyder	*Wireless Telecommunications Networking with ANSI-41, 2/e*

BICSI

Network Design Basics for Cabling Professionals
Networking Technologies for Cabling Professionals
Residential Network Cabling
Telecommunications Cabling Installation

METRO AREA NETWORKING

Metro Area Networking

Steven Shepard

McGraw-Hill

New York   Chicago   San Francisco   Lisbon
London   Madrid   Mexico City   Milan   New Delhi
San Juan   Seoul   Singapore   Sydney   Toronto

The McGraw·Hill Companies

Cataloging-in-Publication Data is on file with the Library of Congress.

Copyright © 2003 by The McGraw-Hill Companies, Inc. All rights reserved.
Printed in the United States of America. Except as permitted under the United
States Copyright Act of 1976, no part of this publication may be reproduced or
distributed in any form or by any means, or stored in a data base or retrieval
system, without the prior written permission of the publisher.

1 2 3 4 5 6 7 8 9 0 DOC/DOC 0 9 8 7 6 5 4 3 2

ISBN 0-07-139914-3

The sponsoring editor for this book was Steve Chapman and the production supervisor
was Pamela Pelton. It was set in New Century Schoolbook by MacAllister Publishing
Services, LLC.

Printed and bound by RR Donnelley.

 This book is printed on recycled, acid-free paper containing a minimum of 50
percent recycled de-inked fiber.

McGraw-Hill books are available at special quantity discounts to use as premiums and
sales promotions, or for use in corporate training programs. For more information,
please write to the Director of Special Sales, Professional Publishing, McGraw-Hill,
Two Penn Plaza, New York, NY 10121-2298. Or contact your local bookstore.

This book is for John Weaver—telecommunications educator, Texan, rodeo fan, restorer of antique chuck wagons, good friend, and a gentleman against whom all others should be measured. Too bad human cloning is prohibited: This world would be a much better place if we had a few more like you around.

CONTENTS

Contents

ACKNOWLEDGMENTS

One more time: I had to dig wide and deep to put this book together. It seems as if every time I had the manuscript ready to go, something substantial would change that required a rewrite. As always, my cadre of editors, critics, and contributors were there for the ride and helped immeasurably. My heartfelt appreciation goes out to Bill Bailey, Paul Bedell, Cyril Berg, Rich Campbell, Phil Cashia, Mark Fei, Marty Flynn, Jack Garrett, Jack Gerrish, Barbara Jorge, Bob Kaphan, Gary Kessler, Phyllis Klees, Naresh Lakhanpal, Gary Martin, Bob Maurer, Richie Parlato, Scottie Ray, Ruth-Anne Renaud, Rick Sanders, Kenn Sato, Kirk Shamberger, Henry Sherwood, Elvia Szymanski, Jack Tongue, Dave Train, Christine Troianello, John Weaver, Ken Wade, Dave Whitmore, Mark Wilson, and Morley Winograd. Thank you for being there with me as you have so many times in the past. I'm sure I've missed a few; I may have forgotten your names here, but I didn't forget your contributions. My thanks and apologies.

Special thanks go out to my friends at Agere Systems, who helped me develop the final section of the book that deals with the components that make it all work. As many times as you've explained it to me, I'm still convinced that there is a high level of magic involved at the component level.

Steve Chapman and Jessica Hornick at McGraw-Hill keep making the good things happen. Beth Brown and Molly Applegate at MacAllister Publishing Services take the output of my pitiful four-finger typing and makes it look like a book. Thanks—nice work.

As always, I reserve this last paragraph for my family. Steve, Cristina, it seems like yesterday you were sitting on my lap listening to Winnie the Pooh; today you're looking at apartments and colleges. Thanks for keeping me (mostly) young. And Bina, thanks for being the person I want to grow old with.

INTRODUCTION

In the last two years, the metropolitan region of the global telecommunications network has come into its own. Once a tiny, almost inconsequential segment that many characterized as "experimental," metro has now become a serious moneymaking contender. A number of factors account for its ascendancy, including changes in the demographic structure of corporations; evolutionary shifts in workflow; the increasing importance of data, information, and knowledge management; and the related importance of database access, changes in network topologies, the rise of Ethernet bandwidth, and the bandwidth discrepancy that exists between the glutted core and the hungry edge. Another key factor is convergence and the evolution of the multiservice network, the greatest use of which appears to be in the metro domain.

There was a time when "corporate headquarters" was exactly that—a single, monolithic building that housed a corporation's employees. Today that environment is evolving such that multiple corporate locations within a metro area are the norm as companies place staff close to major customer clusters. This shift in real estate philosophy, however, brings with it an additional challenge: the need to be interconnected. Furthermore, the network that interconnects these corporate facilities must support a broad range of applications and must satisfy all of their far-ranging *quality of service* (QoS) requirements.

A metropolitan network has the following characteristics. It is found at the edge of the high-bandwidth core network and usually directly touches users as well as the core transport network, serving in effect as the interface point between two distinct regions: access, which defines how users reach it, and transport, which defines how it moves traffic within itself.

The growing demand for bandwidth has driven service providers to invest in an optical infrastructure at the core of the network, and to a lesser degree in the metro and access environments. As a result, the core is now overprovisioned while the edge is massively underprovisioned, resulting in a significant transport disparity.

Optical switching, routing, and multiplexing, as well as unimaginably fast transport, are realities today; unfortunately, they encounter a wall when they try to leave the core and head for the customer over the largely narrowband local loop. This is the focus of the metro technology marketplace. As a result, metropolitan area networking will experience serious growth and interest in the next few years. Companies like Yipes, Cogent, Atrica, and Telseon, which have emerged as metro leaders, now rule this marketplace with such transport solutions as high-speed native Ethernet.

Although numerous technologies lend themselves to metro access and transport, Ethernet has emerged as the clear leader among them. Originally released as a 10 Mbps product based on IEEE standard 802.3 and targeted at the low-end office automation environment, Ethernet rapidly became the most widely deployed *local area network* (LAN) technology in the world. However, as soon as the metro market began to emerge as a "newly discovered network region," it was seen as an alternative transport scheme for both inter- and intrametro network architectures. Today, 10 Mbps is inadequate for many enterprise applications, so new versions of Ethernet have emerged that offer 100 Mbps (Fast Ethernet) 1,000 Mbps (Gigabit Ethernet), and 10 Gbps transport. In keeping with the demands being placed on LANs by the convergence of enterprise applications, standards for LAN-based voice transport that guarantee QoS for mixed traffic types now exist.

Most analysts believe that Gigabit Ethernet will experience a high uptake rate as its popularity climbs. Dataquest predicts that sales will reach $2.5 billion by year-end 2002, a reasonable number considering that 2 million ports were sold in 1999 with expectations of hitting a total installed base of 18 million by the end of 2002. Emerging applications include LAN telephony, *Storage Area Networking* (SAN), server interconnection, and video to the desktop. Various manufacturers have entered the marketplace, including Alcatel, Lucent Technologies, Nortel Networks, and Cisco Systems.

As Ethernet's popularity has grown, many have begun to question the long-term viability of *Synchronous Optical Network* (SONET) and *Synchronous Digital Hierarchy* (SDH) since the two appear to conflict with one another. However, that is not necessarily the case. First, SONET and SDH are physical layer standards that define transmission characteristics for an optical network. Ethernet (actually, IEEE 802.3) is a layer two standard that defines framing and traffic control parameters for a LAN. SONET/SDH have embedded network management and QoS controls, while Ethernet is a best-effort-only framing scheme (although efforts are afoot to add QoS to Ethernet). A number of corollary standards have emerged, including the recently announced IEEE 802.17 *Resilient Packet Ring* (RPR) that will give Ethernet SONET/SDH-like reliability and protection, and effectively extend Ethernet's domain from the LAN to the *metro area network* (MAN) and *wide area network* (WAN). A second standard, ITU X.86, defines an effective mapping scheme for Ethernet frames over SONET/SDH transmission facilities.

As the metro environment has become more important, customer demands have become better defined. First and foremost, customers want a low-cost solution that offers measurable and sustainable QoS and is agnostic with regard to protocols and applications. Second, they want levels of bandwidth that will ensure that blocking and delay are nonproblems, and they want that bandwidth to be scalable. Third, since customers want to keep their costs within bounds, this typically results in the reuse of installed technologies. Finally, they want it to be robust and survivable. So what technologies can satisfy this demand? The answer is not particularly complex; most existing access and transport technologies lend themselves to the metro network domain if they are engineered and deployed properly.

The most common access technologies found in the metro space are, starting with the lowest bandwidth solution, dial, *Integrated Services Digital Network* (ISDN), *Digital Subscriber Line* (DSL), Ethernet, Frame Relay, wireless, and *Asynchronous Transfer Mode* (ATM). Some argue that cable modems should be included because they are beginning to show up in enterprise installations, but they are still minimally deployed in the enterprise and are therefore not yet viable alternatives.

Transport technologies, on the other hand, include *Plesiochronous Digital Hierarchy* (PDH) options (T1/E1, microwave, and so on), SONET/SDH, IP, ATM, freespace optics, and native-mode Ethernet (no collision management).

Metro Network Topologies

What does a metro network look like? Traditionally, metro networks have been based on ring topologies, such as those deployed early on by such first movers as *Metropolitan Fiber Systems* (MFS) and *Teleport Communications Group* (TCG). Today, mesh networks are beginning to appear in the metro because they offer the best characteristics of both: the rapid provisioning and reroute capabilities of a full mesh, and the survivability of the ring. A mesh, after all, is nothing more than a collection of nested rings. Equally valuable is the fact that in mesh architectures, *all* users have access to *all* available bandwidth *all* the time, whereas some ring bandwidth must be reserved for survivability purposes. Rings have another disadvantage as well: The addition of bandwidth in a ring environment requires the addition of a physical infrastructure, an expensive, time-consuming and difficult task, particularly in a city.

The metro environment is a challenging place for service providers to do business. First, installation can be daunting because access to buried conduit is not always available due to location, permit restrictions, or physical crowding. At least one company has come up with the clever idea of running fiber through sewer pipes, a natural (albeit less than pleasant) conduit alternative that has come to be known as "SewerNet." Others have adopted freespace optics, which rely on lasers to carry information between buildings in a city. Another challenge is the *multitenant unit* (MTU). Because a building may host multiple tenants with a variety of connectivity and service requirements, network design and traffic engineering can be complex and never-ending. Finally, management and monitoring is daunting because of the topology of the network and the variety of customers using it.

The Metro Marketplace

The metro marketplace has four layers. The lowest, and like all food chains the layer upon which all others depend, is the component manufacturer. Directly above all the layers is the system manufacturer. Finally, near the top, we find the service provider. The fourth layer represents the customer or user of metro network technologies.

The component manufacturers face a number of serious challenges, not the least of which is the need to support an extremely broad array of service interfaces to satisfy their system manufacturer customers. They must anticipate requirements in multiple global market areas, monitor the status of emerging standards and technologies, and stay abreast of evolving customer application demands that will ultimately feed into the design process for the devices they manufacture for the higher layers of the food chain.

The system manufacturers, including Lucent, Nortel, Cisco, Fujitsu, and a host of others, are equally challenged but at a different level. They must monitor end-user applications to ensure that the devices they manufacture address evolving and constantly changing functional requirements. They must also be constantly aware of service provider office engineering issues, including footprints, heat dissipation, power consumption, and so on.

The service providers must look in both directions. They must look up toward their enterprise customers and those customers' customers who will benefit from the services delivered by the provider. They must also be ever cognizant of their own service restrictions and capabilities as well as those of the many hardware vendors they purchase from.

Finally, we have the customers. They are the driving force behind the efforts of the other three layers and must therefore be represented here. Their applications and bandwidth demands set the tone for the other three layers of the metro networking food chain.

Conclusion

The metropolitan network has grown from being an incidental area of the network to one of the most important and revenue-rich technology segments today. It is a highly innovative environment, combining the best features of both legacy and emerging technologies. It is also an area of significant opportunity—and peril—for players at all layers of the metro technology food chain. Readers, whether sales, engineering, design, or marketing professionals, or simply interested bystanders, must be keenly aware of not only the technologies found in metro networks, but the topologies deployed, the interfaces supported, the applications that drive it all, the regulatory considerations, and the economics involved. A brilliant spotlight has been focused on the metro domain today, and any number of players are wrestling over who gets to control that spotlight, including regulators, end users, bandwidth providers, and application developers. As my friend and colleague Dave Train likes to observe, the metro marketplace is the "next" final frontier. He's absolutely right: The transport world is well established and evolves apace. Access, while still evolving toward broadband critical mass, is well understood and will continue to enjoy attention because it touches the customer. Metro, however, is the technological land of opportunity, and the players are lining up to stake claims in what will be an enormous field of play.

About the Book

I wrote this book because it is clear to me that as Dave Train says, metro is the "next" final frontier. Like all new areas of interest, metro networks are pushing the envelope of innovation and promise and are forming the basis for a whole new set of applications and capabilities. They are also giving new life to old technologies such as SONET/SDH and Ethernet.

One reason I find the metro network world so interesting is this: As the need for multiservice metropolitan networks has grown, both old and new technologies have stepped into the breech to address the problems stated by customers. Metro networks, then, have evolved in lockstep with the customers that depend on them. As a result, they actually work: They address specific concerns and challenges and do so in innovative, cost-effective, and efficient ways. They do not favor new whiz-bang solutions simply because they are new whiz-bang solutions; instead they pick whatever technology best addresses the need at hand. If that solution happens to be a legacy technology, then so be it—as long as it fixes the problem. Similarly, if something new is a better answer, then it finds a home.

Metro Area Networking comprises five parts. Part 1, "A Lesson in Medieval History," provides a general overview of networks and defines the metro network by placing geographic and technical bounds around its margins. Included are discussions of major industry trends, competing topologies, and an introduction to the evolution of networks.

In Part 2, "The Metro Network," we describe the evolution of metropolitan networks and the drivers behind their evolution. Included are discussions of regulatory, economic, demographic, and technological factors.

Part 3, "Enabling Technologies," provides an overview of the technologies found in metro networks. It breaks them into three groupings: legacy solutions, transitional technologies, and future solutions.

Part 4, "Metro Applications," describes metro applications, including data mining, information and knowledge management, *Customer Relationship Management* (CRM), *Enterprise Resource Planning* (ERP), the various storage solutions that make them possible within the metro domain, and, finally, network management.

Finally, Part 5, "Players in the Metro Game," provides a survey of metro players, taking into account the four market regions that empower the metro environment: the semiconductor and opto-electronic manufacturers, the system manufacturers, the service providers, and, of course, the end users.

As always, I enjoy hearing from readers, so please feel free to contact me with suggestions that would improve the book, or if you simply want to chat about this wonderful industry we are privileged to be part of. The danger of writing a book about something as new as metro networking is that as the environment changes, some parts of the book become dated quicker than others. So please let me know if you run across something while reading that needs to be updated. I'll see to it that the change makes it to the next edition. Thank you.

Steven Shepard
March 2002
Williston, Vermont; Beijing, China; and Newport Beach, California
Steve@ShepardComm.com

A Lesson in Medieval History

It is fascinating to me how much my interest in history helps me understand the evolution of modern networks and the inner workings of the technologies that power them. Metro networks are no exception. As I studied the evolution of the network from a relatively monolithic transport construct to a more differentiated and specialized set of technologies for performing access and transport tasks, I began to see parallels between it and —of all things—the development of medieval village societies. I had recently read Frances and Joseph Gies' *Life in a Medieval Village* and *Life in a Medieval Castle,* and many of the things I learned while reading them were still rattling around in my head.

When formal societies were still coalescing in Europe—prior to the sixth century or so—the world was a forbidding, often dangerous place. People huddled together in massed groups for protection because no form of social order existed to create formal, organized protection, division of labor, or any of the other characteristics of a modern society.

As time passed, the social order evolved. The feudal castle model emerged in which people lived behind the walls of a fortress and relied on an organized militia for protection. Some people lived and worked outside the walls, but they were agricultural laborers who needed to interact with the land. The artisans and craftspeople (the more specialized laborers) lived within the castle that belonged to the landholder.

The next evolutionary stage saw the development of the medieval village in the tenth century or thereabouts. As the Gies observe in *Life in a Medieval Village*[1],

> The medieval village . . . was the primary community to which its people belonged for all life's purposes. There they lived, there they labored, there they socialized, loved, married, brewed and drank ale, sinned, went to church, paid fines, had children in and out of wedlock, borrowed and lent money, tools, and grain, quarreled and fought, and got sick and died. Together they formed an integrated whole, a permanent community organized for agricultural production.

Life, then, was fairly concentrated in the sense that few strayed far from the fold. The only exceptions were those who struck out on their own to explore new frontiers, to expand the knowledge that society had at the time of life beyond known borders. They went in search of new opportunities and wealth, and often found it. However, they typically had to be supplied by caravans that traveled long distances to reach them, often at great peril.

---

[1] *Life in a Medieval Village*, page 7.

Over time, societies matured and became more complex. Instead of everyone living in a geographically constrained area, the boundaries of the societal model began to expand as new villages emerged. As social order arrived and stayed, and as a social fabric was woven that defined the mores of behavior for the people who lived in the society, the world became marginally safer. Soon people were straying farther from home and establishing cultural beachheads, and modern Europe began to flourish.

From a highly centralized model, society evolved to include various modalities: the central castle that protected royalty and the wealthy merchant class, a collection of agrarian workers living immediately outside the castle and providing food to those within in exchange for protection, and a network of hamlets and villages scattered at the periphery of the territory overseen by the barony of the castle. How interesting it is to note that in many ways modern voice and data networks took a remarkably similar developmental path.

The Evolution of Modern Networks

The first networks, of course, were simple point-to-point connections that interconnected individual households or businesses. The concept of switching had not been developed yet, so anyone who wanted to communicate with anyone else had to have a dedicated circuit between the two locations. There was a time when AT&T tried to gain control of rooftops in cities so that they would have a place to install the myriad wire pairs required to interconnect everyone that wanted service. Soon the skies were filled with wire, as shown in Figures 1-1 and 1-2. The only way to provide universal connectivity was to have individual connections for every conversation, and that required a *lot* of wire. Technically, a network that enabled any-to-any connectivity had to be fully meshed, as shown in Figure 1-3.

With the arrival of switching came the ability to offer connectivity on demand. Instead of dedicated connections between every pair of would-be callers, each subscriber needed a connection from his or her house or business to the telephone company central office. There the switch (originally it was a human switch—the operator—shown in Figure 1-4) would take the call setup information from the subscriber and create a temporary connection between the calling and called parties. The result is that the bulk of the resources required to support telephony migrated into the core of the telephone network because it made technical and economic sense to do it that way.

Figure 1-1
Aerial cable in a
downtown area
(Source: Lucent
Technologies)

Figure 1-2
The number of
crossbars on this pole
illustrates the amount
of wire that had to
be in place to
support telephony in
the days before
switching.
(Source:Lucent
Technologies)

As switching evolved and the human operator was replaced first by the mechanical, then the electromechanical, and finally the all-electronic switch, the intelligence that was part of the switching infrastructure began to show its value to both the service provider and the subscriber. New appli-

Figure 1-3
A fully meshed network, showing any-to-any connectivity

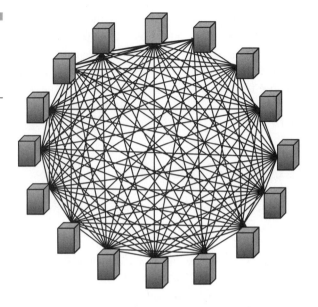

Figure 1-4
Operators sitting at an old-style cord board
(Source: Lucent Technologies)

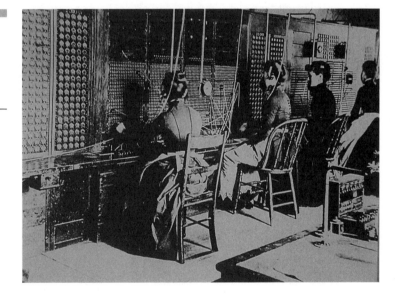

cations evolved based on the marriage of computer and networking technologies, and the so-called intelligent network was born. This, in turn, led to the development of sophisticated capabilities that added to the perceived value of the network, which were the famous value-added services, including Caller ID, three-way calling, call return, and call trace.

As the network grew and became more capable, it also became more functionally diversified. The voice network evolved into local and long-distance domains, helped along by a series of regulatory decisions that forced a functional and business-based schism between the two. Local service became primarily the access arm of the business, while long distance provided connectivity between far-flung communities.

Further Diversification

In the 1970s, a new commercial phenomenon emerged: data transport. Corporations began to develop *management information systems* (MISs) and to forge new organizations to manage them, such as the *Information Systems Organization* (ISO or IS). Deep within the viscera of corporate data centers (see Figures 1-5 and 1-6), the mainframes and minicomputers, vital organs of the body of information that gave corporations their competitive edge, digested and processed the raw data they collected from the never-ending stream of transactions that fed them. These massive, centralized repositories relied on a web of dedicated, private-line facilities to interconnect them to the service centers, billing offices, and customer record centers that generated the traffic they managed.

In the early 1980s, the PC arrived, and with it came a major reinvention of the data phenomenon. Instead of relying exclusively on centralized data-processing facilities, corporations began to rely more on a distributed pro-

Figure 1-5
Mainframe
computers in a
modern data center

Figure 1-6
A tape cartridge library in a modern data center

cessing model, which meant that like a cow with its five stomachs, the data was already somewhat digested by the time it arrived at the data center for processing. Instead of private-line circuits offering dedicated connectivity for the now famous "dumb terminals" through some kind of terminal multiplexer, PCs replaced the dumb devices. This began the migration of computing intelligence from the core of the network to the edge, where the users could take advantage of new capabilities to become better, more efficient workers.

During this evolution of computer usage, the network went through a similar morphogenesis. As mentioned earlier, it initially had two regions: access and long-haul transport. Over time, however, a third region—the customer environment—was added to the list. This was where the *local area network* (LAN) first emerged, as did Ethernet, the most common of all LAN protocols. Thus, the customer environment (the LAN) generated and terminated traffic, the access network connected the customer to the network, and the transport network provided long-haul connectivity between customer locations. Interesting, isn't it, how this compares to the castle, roadways, hamlets, villages, and far-flung outposts of early Europe?

Core Versus Edge: A Few Words

Network-edge devices typically operate at the "frontier" of the network, serving as vital service outposts for their users. Their responsibilities typically include traffic concentration, which is the process of statistically balancing the load against available network resources; discrimination, during which the characteristics of various traffic types are determined; policy enforcement, the process of ensuring that required *quality of service* (QoS) levels are available; and protocol internetworking in heterogeneous networks. Edge devices are often the origination point for *Internet protocol* (IP) services and typically provide no more than 20 Gbps of bandwidth across their backplanes.

Core devices, on the other hand, are responsible for the high-speed forwarding of packet flows from network sources to network destinations. These devices respond to directions from the edge and ensure that resources are available across the *wide area network* (WAN), and guarantee that QoS is guaranteed on an end-to-end basis. They tend to be more robust devices than their edge counterparts and typically have 20 Gbps or more of full-duplex bandwidth across their backplanes. They are nonblocking and support larger numbers of high-speed interfaces.

As the network has evolved to this edge/core dichotomy, the market has evolved as well. As convergence continues to advance and multiprotocol, multimedia networks become the rule rather than the exception, sales will grow exponentially. By 2003, consultancy RHK estimates that the edge switch and router market will exceed $21 billion, while the core market will be nearly $16 billion. At the core, Cisco currently holds the bulk of the market at roughly 50 percent, slightly less at the edge with 31 percent. Other major players include Lucent Technologies, Marconi, Nortel Networks, Juniper, Newbridge, Fore, Avici, and a host of smaller players.

It is interesting to note that in order to adequately implement convergence, the network must undergo a form of *divergence* as it is redesigned in response to consumer demands. As previously described, the traditional network concentrates its bandwidth and intelligence in the core. The evolving network has in many ways been inverted, moving the intelligence and traffic-handling responsibilities out to the user, replacing them with the high-bandwidth core described earlier. In effect, the network becomes something of a high-tech donut. A typical edge-core network is shown in Figure 1-7.

The core, then, becomes the domain of optical networking at its best, offering massively scalable bandwidth through routers capable of handling both high volume traffic and carrying out the QoS dictates of the edge devices that originate the traffic.

The drivers behind this technology schism are similar to those cited earlier. They include

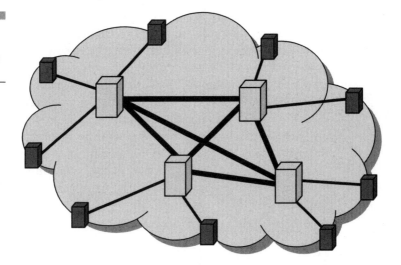

Figure 1-7
The edge and core regions of a modern data network

- The need to create routes on demand between individual users as well as between disparate work groups in response to the market shying away from dedicated, costly facilities.

- Guaranteed interoperability between disparate protocols.

- Universal, seamless connectivity between far-flung corporate locations.

- The optimum utilization of network bandwidth through the appropriate use of intelligent prioritization and routing techniques.

- Traffic aggregation for wide area transport to ensure the efficient use of network bandwidth.

- Granular QoS control through effective policy and queue management techniques.

- The growing deployment of high-speed access technologies such as *Digital Subscriber Line* (DSL), cable modems, wireless local loop, and satellite connectivity.

And why is this evolution occurring? Because the closer the services a network provider sells are placed relative to the customer, the more customized, targeted, and immediate those services become. When they are housed in a shared central office, they are much more generalized, catering more to the requirements of the masses and treating the customer as if their requirements were commodities. As the network evolves and a clear functional delineation between the edge and the core becomes visible, the role of the central office suddenly changes. In fact, the central office largely disappears. Instead of a massive centralized infrastructure from which all services are delivered (similar to the model employed in legacy data centers), we now see the deployment of hundreds of smaller "regional offices" placed close to customer concentrations and housing the edge-switching and routing technology that deliver the discrete services to each customer on an as-required basis. Those smaller offices are in turn connected to the newly evolved optical core that replaces the legacy central office and delivers high-speed transport for traffic to and from the customers connected to the edge offices. This is the network model of the future: It is more efficient and places both the services and the bandwidth that they require where they belong.

Of course, the network management model must now change in response to the new network architecture. Instead of managing centralized network elements—a relatively simple task—the management system must now manage in a distributed environment. This is more complex, but if done properly it results in far better customer service because of the immediacy that results from managing at the customer level.

Customer applications have evolved to require more bandwidth in response to its enhanced availability at a reasonable price, and dramatic augmentations have been made to the capabilities of *customer-provided equipment* (CPE), such as bridges, routers, switches, and multiplexers. Because of these factors, the traditional requirement that the aggregation, prioritization, and concentration of traffic be done in the network core is no longer valid. Additionally, network protocol models have evolved to the point that traffic prioritization and QoS guarantees no longer have to derive from central-office-based hardware and software. In fact, more and more, modern networks have evolved such that they now comprise two primary layers: an outer edge region, responsible for QoS policy enforcement, and a central core region, responsible for carrying out the mandates of edge devices and transporting traffic at all possible haste across the WAN. One driver behind this evolution is the shift in protocol models as IP-based networks become more and more common, and IP delivers upon its promise of a protocol-agnostic, high-bandwidth, QoS-driven architecture.

The Birth of Metro

In the late 1980s, a new network region began to emerge. Driven by growing bandwidth requirements in campus installations and to a certain degree by the mixed bandwidth demands of multitenant office buildings in major cities, the first rudimentary form of *metropolitan area networks* (MANs) arrived.

There were good reasons for this early evolution. Corporations were beginning to rely more and more on their corporate data centers and the services they provided for competitive advantages. As a result, they also relied heavily on network connectivity to provide access to databases from remote locations. Traditional private-line technologies were neither cost-effective nor redundant, and the cost to add redundancy was prohibitive. When the first metro carriers (Teleport Communications Group and Metropolitan Fiber Systems, the dreaded bypassers) came along with promises of optical connectivity, corporate *information technologies* (IT) and telecom managers sat up and took notice. *Synchronous Optical Network* (SONET) in the United States and *Synchronous Digital Hierarchy* (SDH) in Europe and the rest of the world were emerging as powerful ring-based transport options, and these companies built major installations in large cities based on these technologies. With promises of dual entry and 24×7 monitoring, they cornered a large segment of the enterprise market.

The next technologies to step into the spotlight were *Fiber Distributed Data Interface* (FDDI) and the *Switched Multimegabit Data Service* (SMDS). FDDI was really something of a cross between the ring structure of SONET and IBM's Token Ring, while SMDS was for all intents and purposes a form of connectionless *Asynchronous Transfer Mode* (ATM). Both were designed to offer high-speed service within the metro environment, and although they did so quite effectively, neither of them enjoyed significant longevity. There were several reasons for this: one was the fact that the metro market was new and really had not achieved critical mass as yet; another was that the demand for bandwidth had not yet reached the levels that either was capable of delivering; and the third was that both FDDI and SMDS were quite complex.

Fiber Distributed Data Interface (FDDI)

FDDI was designed to address the growing requirements for connectivity in the metropolitan area. It had two key applications: to serve as a premises-wide backbone for LAN traffic in business parks or other campus environments, and to provide connectivity between high-speed workstations at research institutions.

FDDI relies on token passing to control access to the shared transmission medium. First, it is important to note that token passing is typically implemented on ring networks. As a result, traffic always travels in the same direction, and the order of appearance of the stations on the LAN never changes. One consequence of this is a very predictable *maximum token rotation time*, which translates to a bounded delay in the transmission of data. This becomes a significant factor when latency is an issue.

When a station wants to transmit a file to another station, it must first wait for the "token," a small and unique piece of code that must be "held" by a station to validate the frame of data that is created and transmitted.

Let's assume for a moment that a station has secured a token because a prior station has released it. The station places the token in the appropriate field of the frame it builds, adds the data to be transmitted along with the destination address information, and transmits the frame to the next station on the ring. The next station, which also has a frame it wants to send, receives the frame, notes that it is not the intended recipient, and also notes that the token is busy. It does not transmit but instead passes the frame of data from the first station on to the next station. This process continues, station by station, until the frame arrives at the intended recipient on the ring. The recipient validates that it is the intended recipient, at which time it makes a copy of the received frame, sets a bit in the frame to indicate that it has been successfully received, *leaves the token set as busy,* and transmits the frame on to the next station on the ring. Because the token is still shown as busy, no other station can transmit. Ultimately, the frame returns to the originator, at which time it is recognized as having been received correctly. The station removes the frame from the ring, frees the token, and passes it on.

FDDI differs from traditional token ring in one very significant way: It has two rings, not one. FDDI is a dual, counter-rotating optical ring architecture that operates at 100 Mbps and is capable of supporting as many as 500 workstations with distances between nodes as great as 2 kilometers. In a dual-ring deployment, the total ring length cannot exceed 100 kilometers.

The dual-ring topology, shown in Figure 1-8, is primarily for fault tolerance. For the most part, stations are directly connected to both rings. One ring operates in a clockwise direction, while the other transmits data on the counterclockwise ring. One ring can be used as the primary path while the other can be used as a back ring. If the primary ring fails, stations on either side of the fiber cut realize that the signal has been lost, and the backup ring becomes the primary data path. In the event that both rings are cut, the stations on either side of the breach "wrap" internally, as shown in Figure 1-9.

Figure 1-8
The typical dual-ring architecture found in optical networks

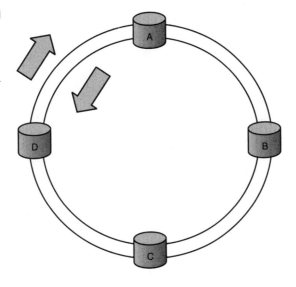

Figure 1-9
The failure of both rings causes the ring to wrap, ensuring connectivity.

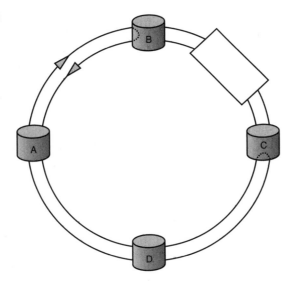

Stations can be attached to FDDI rings in a variety of ways, as shown in Figure 1-10. A *single attachment station* (SAS) is only connected to the primary ring, while a *dual-attached station* (DAS) is connected to both the primary and backup rings. An alternative technique is to use what is called a *dual attachment concentrator* (DAC) that attaches to the rings and serves

Figure 1-10
A representative FDDI
ring showing SASs,
DASs, and a DAC
with multiple stations

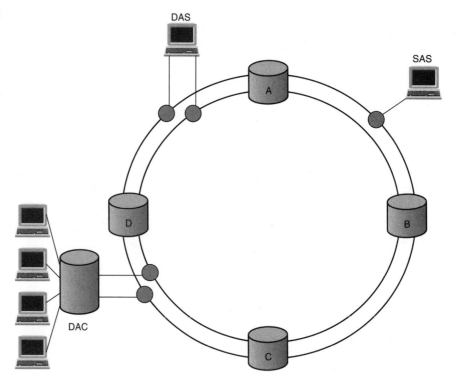

as a multiplexer of sorts, enabling multiple stations to access the rings as required.

This architecture has pros and cons. It provides very good fault tolerance since the workstations themselves are isolated from the ring and therefore cannot disrupt transmission, should an individual workstation fail. It also enables easy network planning and lower-cost deployment. Furthermore, the technique is less disruptive to the overall network since the ring is not affected when stations are added or removed. On the other hand, the concentrator represents a failure point that could conceivably affect multiple stations.

FDDI is still used in some corporations, but the rise of Fast and Gigabit Ethernet has for the most part stymied its proliferation. The other technology that heralded the arrival of the metro network was SMDS.

Switched Multimegabit Data Service (SMDS)

SMDS, largely gone now, was a connectionless, high-speed, packet-based data transport mechanism. It supported standard protocols, interfaces, and technologies.

SMDS enabled users to extend their data communications reach over a broad geographical area. Since it was a service offered by the telcos rather than a technology, it permitted this distance extension using existing *Customer Premises Equipment* (CPE) with minimal upward investments in leased lines as the number of users increased.

SMDS was defined by the *Institute of Electrical and Electronics Engineers* (IEEE) in its 802.6 MAN standard and was adopted and implemented by Bellcore, now a division of the *Science Applications International Corporation* (SAIC) known as Telcordia.

The technology's evolution occurred in parallel with ATM, and because SMDS is also a cell-based service, it enjoyed quite a bit of attention as ATM's ascendancy began. Both services rely on 53-octet cells and can accommodate packet lengths as large as 9,188 octets (the maximum packet size for SMDS is 9,188 octets, while the maximum length for ATM is much larger.

SMDS relies on three distinct protocols for its general functionality: *Distributed Queue, Dual Bus* (DQDB), the *SMDS Interface Protocol* (SIP), and the *Data Exchange Interface* (DXI). DQDB defines the manner in which stations gain access to the shared medium, SIP defines the interface between the end user and the medium, and DXI defines a standard protocol at the physical and data link layers for communication between the router and the *data service unit* (DSU)/*channel service unit* (CSU). Most North American implementations relied on DQDB, with DS1 or DS3 access for connectivity.

SMDS was designed to provide broadband services across the metro region, but because it was a public offering it never enjoyed the uptake rate that would enable it to reach critical market mass. Banks, for example, expressed concerns over the transport of private banking data over a public network infrastructure, even though there was no justification for the concern. *Virtual private networks* (VPNs) with secure encryption protocols and tunneling capabilities had not yet been invented, so no precedent had been established to define just how secure a public infrastructure solution could actually be. Many industry pundits claim that if SMDS had simply been renamed "Connectionless ATM," the service providers would not have been able to keep up with orders. Unfortunately, like FDDI, it withered away and has for the most part disappeared. The metro market, of course, has continued to expand as new technologies have stepped into the breech. One of these, Ethernet, has been "reinvented" and is succeeding handily. More on Ethernet later.

The Maturing Metro Market

According to a recent study by consultancy RHK, the unstoppable demand for bandwidth in North America will generate $4 billion in new revenues by 2006 for service providers that offer optical Ethernet connectivity within the metro space. Similarly, In-Stat predicts that the metro optical network market will grow from an already respectable $13 billion in 2001 to $23.6 billion by 2005, which moves the market segment to the top (or at least very close to the top) of the list of high-growth telecom segments. An array of new technologies is creating deployment strategies that meet the needs of metropolitan optical networks by not only introducing new technology solutions, but by also extending the life of installed legacy plant and protocols that are perfectly serviceable and may not yet be fully depreciated.

The current state of the economy has caused service providers and other formerly capital-rich companies to reconsider their spending strategies. Ethernet and other legacy solutions are well understood, cost effective, and, given their current operational characteristics, perfectly capable of supporting the demand for broadband access and transport that is growing apace in the metro marketplace. SONET and SDH now routinely operate at 10 Gbps, and some next-generation optical hardware efficiently manages packet traffic and provides support for the burgeoning multiservices network. *Dense Wavelength Division Multiplexing* (DWDM) systems are being deployed in both core transport and edge access environments, and Optical Ethernet, operating at 1 Gbps and higher, is blurring the line between the LAN and the MAN.

Metro Market Segments

Today, the metro marketplace has changed dramatically. It has come into its own and become the focal point for a significant number of companies that include component manufacturers, system manufacturers, service providers, and the end users within the metro marketplace (see Figure 1-11). In fact, these four segments make up something of a food chain in the metro arena: End users buy services from the service providers, service providers buy systems from the systems manufacturers, and systems manufacturers buy semiconductor and opto-electronic components from the component manufacturers.

In reality, the metro networking environment has become extremely important to all four layers of the food chain. The reasons are well known, but it would be wise to review them here.

The first reason was stated earlier: Metro is the last frontier in the development of the global telecommunications network. Because it lies close to the customer, and because it serves a *new* type of customer (the multiservice enterprise metro user), untold revenues can be mined from the segment. Keep in mind that the greatest diversity of new opportunities always lies at the periphery of the current operating domain, regardless of whether the discussion topic is networks, computers, biotechnology, or European explorers pushing the limits of geographic knowledge.

We often hear tales about the optical glut in today's network. Let us be perfectly clear about the truth behind that observation: Although an overabundance of optical fiber has been installed in the core transport network,

Figure 1-11
The metro
marketplace food
chain

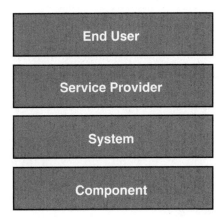

and therefore an attendant overabundance of available bandwidth exists, *there is a dramatic paucity of bandwidth at the edge*. Broadband access is still not as widely available as customers and service providers would like, and the metro space is equally underrepresented. The primary reason for the lack of bandwidth at the edge is regulatory, not technological in nature, and although regulatory bodies are now coming to grips with the results of earlier decisions and taking steps to move broadband forward, such things take time and are politically charged. We'll talk more about the impact of regulatory decisions later in the book.

Another driving force is service creation: The metro network lies between the customer-premise-based LAN and the service provider's transport network. It therefore plays a significant part in the service provider's ability to offer what customers demand, particularly given the dynamic nature of the metro networking environment. As LAN solutions become more service-capable and as transport technologies plummet toward commodity status, the need to bridge what appears to be an "unbridgeable gap" between the service-rich edge and the high-bandwidth, low-cost core becomes critically important to both service providers looking to preserve their dwindling revenue base and customers looking for diverse service capabilities.

Of course, like anything new, the metro network environment has its set of challenges that present difficulties for service providers and users alike. The metro domain is neither local, access, or transport; in fact, it has elements of all three. It is neither circuit-switched nor is it packet-based; instead, it's a hybrid of the two. Finally, metro networks are not well understood yet, which means that the enormous volumes of data that they have now been found to generate are not necessarily well served by current network designs. The diverse collection of applications found in the metro—LAN traffic, video, high-bandwidth data, voice, Internet, and so on—makes capacity design considerations something of an arcane art. Most metro infrastructures are built upon legacy SONET rings, which are fine for the transport of synchronous, high-speed data generated by traditional applications, but they are inadequate for the requirements of less predictable bursty traffic streams from packet-based applications. And when the two are mixed across the same network, one typically dominates network resources and somebody's service quality suffers.

Another issue is protocol complexity. Consider the requirements of the typical enterprise network customer: They want (at a minimum) low-cost, high-bandwidth transport; 24/7 proactive network management and vendor support; discrete, configurable, dynamic, and measurable QoS with an attendant *service level agreement* (SLA); and multiservice, multiprotocol support. To accomplish this today, networks look like a layer cake: multichannel DWDM at the lowest level for bandwidth; SONET or SDH at the next level for management and international standardization; ATM at the next level for predictable, configurable, and measurable QoS; and IP at the highest layer for multiprotocol support. The protocol overhead alone is out of bounds. SONET reserves roughly 5 percent of each frame for management overhead, ATM reserves 10 percent of every cell for network overhead, and even IP holds back a small percentage of the available bandwidth for packet overhead. If the protocol complexity can be reduced, and if some of the overhead can be eliminated or at least minimized, a far more efficient network will result.

Consider the diagram shown in Figure 1-12. The left side of the illustration details the model we have just described: IP over ATM over SONET/SDH over DWDM. This model is by far the most costly, most complex, and least efficient protocol stack that could be implemented. It does the job, however. All the requirements (high bandwidth, low cost, QoS, multiprotocol support) are possible with this four-tier model. It represents the legacy model of service transport.

Over time, one plan will evolve through the two possible scenarios shown in the center of the diagram. The first one, IP over ATM over DWDM, assumes that SONET and SDH are no longer required because the management and framing they typically offer are being done somewhere else, thus reducing one layer of overhead. Similarly, the next transitional model assumes that ATM has been eliminated, presumably because its granular QoS capabilities are being performed by another protocol, presumably *Multiprotocol Label Switching* (MPLS) or *Generalized MPLS* (GMPLS) (more on these later in the book). With this model, the ATM "cell tax" is eliminated. Finally, by ultimately implementing the model shown on the right

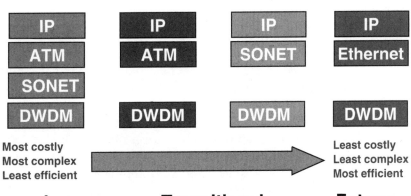

Figure 1-12

The evolution of technology in the metro network

Legacy.........Transitional............Future

side of the illustration, we eliminate both ATM and SONET/SDH. IP packets are instead encapsulated in Ethernet frames for transport across the network, and DWDM channelization provides all the necessary bandwidth. This is clearly the most cost-effective, efficient, and least complex model available to network providers and customers today. And although it has not yet been widely implemented, it will be as networks evolve toward their future architecture schemes.

As a result of all these complex issues, the metro network is a complex environment to work within and is therefore a challenge to both manage and budget. Clearly, the metro space is not going to go away. It's not one of those passing fads, nor is it a network region that is in danger of being subsumed by another. So what efforts are underway to help overcome the challenges posed?

First of all, given the growing influence of Ethernet in all its many forms and the perception that SONET and SDH are inflexible and getting a bit long in the tooth for certain applications, a number of standards bodies have kicked off efforts to reexamine and redefine the standards that govern goings-on within the metro arena. Among these are IEEE efforts for Gigabit and 10-Gigabit Ethernet; ITU X.86, which maps Ethernet frames into SONET/SDH frames; and the IEEE's 802.17 standard for the recently announced *Resilient Packet Ring* (RPR) technology that will ultimately replace SONET and SDH in certain situations. Other standards work is just beginning on such technology areas of interest as Carrier-Class Ethernet (Ethernet with QoS so that it can offer the legendary "five nines of reliability"), switching infrastructures based on MPLS and/or GMPLS, the long-term use of DWDM, the role of next-generation intelligent networks, and the evolution of network management requirements.

Carrier-Class Ethernet

Carrier-Class Ethernet has enjoyed significant attention in recent months for all the right reasons. It is well established and understood, it is user-friendly, it is internationally standardized and accepted, and it is inexpensive. Because of these factors and the evolving demands of the marketplace, Ethernet is finding itself thrust in a variety of directions. It is seen as a potential replacement for SONET and SDH, and as a service interface capable of supporting multiprotocol/multiservice access for such protocols as traditional voice, *voice over IP* (VoIP), VPNs, Frame Relay, ATM, and IP.

Traditional Ethernet was never intended to address the applications it is currently being positioned to resolve. However, the arrival of Gigabit and 10- Gigabit Ethernet has changed that, and the technology is taking on new responsibilities for which it is imminently qualified. For example, 10-Gig Ethernet will be formally ratified later in 2002; already, however, a number of 10-Gig Ethernet solutions are on the market.

Ten-Gigabit Ethernet is quite flexible and can be implemented in a range of topologies, including connection-oriented networks, traditional Ethernet, switched networks, mesh topologies, and rings. It can be deployed in point-to-point architectures or as an aggregation technology for lower-speed Ethernet segments.

The technology has a variety of advantages that include scalable bandwidth ranging from 64 Kbps to 1 Gbps, the capability to offer service-level granularity (including burst and bit-rate guarantees), rapid provisioning, point-to-point and point-to-multipoint installation options, and compatibility with legacy SONET installations through the transmapping of X.86.

Of course, like any new technology, Carrier-Class Ethernet has its share of detractors, many of them among the carriers. Their single greatest concern is the degree to which Ethernet can handle both bursty data applications and the synchronous requirements of voice and continuous data. They are also concerned about the technology's capability to support high-speed restoration and rerouting as well as legacy solutions already do. Some believe that although Ethernet is highly capable, it is not the most efficient protocol available for existing applications. Some believe that it is wasteful of bandwidth when implemented in ring configurations, the primary reason that IEEE 802.17 (RPR) is under development. As we will see later in the book, RPR is extremely efficient, making the most of available ring bandwidth and offering 50-millisecond protection switching.

MPLS/GMPLS-Enabled Networks

MPLS, first created by Cisco under the name tag switching and later adopted by the *Internet Engineering Task Force* (IETF), combines data-link-layer information about the state of network links such as bandwidth, latency, and overall utilization with network-layer information (typically IP) to improve the efficiency of packet exchange. It gives service providers significant flexibility for traffic rerouting in order to avoid link failures, congestion, and network bottlenecks.

One of the greatest beneficiaries of MPLS is *Internet service providers* (ISPs), who will be able to manage different kinds of traffic based on priority and requested QoS levels. This translates into variable billing options, which in turn translates into adaptable customer service.

In MPLS networks, *label edge routers* (LERs) assign a label (in reality, an identifier) to each packet. These labels contain destination information, bandwidth requirements, delay limits, and a variety of other measurements. Once the packets have been classified, they are assigned to *label switched paths* (LSPs), where *label switched routers* (LSRs) assign labels to the outgoing packets. As a result, network operators can reroute traffic based on a variety of customer requirements and characteristics. Packet processing occurs much faster because packets do not have to be opened and examined by every router along the path.

GMPLS, on the other hand, is designed to enable edge devices such as routers and switches to request bandwidth directly from the optical layer. It is also designed to make the optical network more dynamic by migrating intelligence into some traditionally "dumb" optical elements such as optical cross-connects. By adding intelligence in specific areas of the optical network, providers can offer faster restoration of service following network disruptions and more efficient bandwidth utilization.

GMPLS also will deliver information about the optical network to embedded routers, enabling them to make more informed decisions about network routing using knowledge about the "health" of routes through the optical network. GMPLS lends itself to the management of *Time Division Multiplexing* (TDM)-switched networks, packet-switched networks, lambda-switched optical networks, and fiber-based transport. In short, it attempts to inject the sophistication of the router into the optical layer for a wide variety of network types.

DWDM

Conduit space in the metropolitan area has been depleted in many cities, and the cost to add more can be prohibitive, as can the morass of permits and bureaucratic paperwork required to secure permission to dig up city streets. DWDM is one alternative that has received a great deal of attention. In the long-haul environment, the cost to deploy additional fiber as a solution to capacity exhaust is approximately $70,000 per fiber mile, compared to $12,000 for the same capacity addition accomplished through DWDM deployment. In the metro, those numbers are higher because of the environment: more congestion, more traffic, and more obstacles. Nevertheless, DWDM is a good solution because it offers the opportunity to add more available bandwidth by changing the endpoint electronics rather than trenching and physically adding fiber.

Metro DWDM is the latest arrival on the scene, with devices offering as many as 32 wavelengths per fiber at 10 Gbps per channel. Forty Gbps is just around the corner, which adds more credence to the claim that it is far more cost effective to replace the endpoint electronics than it is to physically add fiber. Of course, this adds a whole new management element that to date has largely been ignored.

Bandwidth management at the optical level has traditionally been primitive at best, a serious problem for the metro service provider that wants to offer differentiated services based on discrete optical network management. The result of this is the arrival of a whole new set of companies in the optical space that offers products known as *multiservice provisioning platforms* (MSPP). These platforms enable the service provider to manage their infrastructure components and service offerings with far more granularity. They offer service providers the ability to gradually migrate to an all-optical infrastructure when the time arrives, to support wavelength-based services at the optical level, and to support legacy, transitional, and future services all on a common network infrastructure. This is a powerful capability indeed. Lucent Technologies was one of the first companies to offer this service with its WaveWrapper technology, and others have followed. A number of new and significant standards have been put into place in the last couple of years that address wavelength management, including G.959.1 and G.709.

Many vendors are taking this approach, using proprietary methods to perform optical framing and wrapping. In other words, they encapsulate client signals in a framing scheme—using digital wrappers—that supports management and monitoring functions as well as higher-order multiplexing. The purpose of optical framing in DWDM systems is to produce a wavelength-aware device that can provide transport for all the other major lower-layer protocols, including SONET and SDH.

These techniques are proprietary now, but standards for DWDM-enhanced/all-optical networks are falling into place. A number of optical network standards were presented during 2001, including G.709 and G.959.1, which address the physical-layer interfaces and the frame format for all-optical networks. G.709 was carefully written to incorporate many of SONET and SDH's more valuable characteristics, including performance monitoring and span protection, yet it also adds *forward error correction* (FEC) and end-to-end wavelength management (to address the arrival and incorporation of DWDM). The only disadvantage of G.709 is that it requires a substantial investment in new network infrastructure; the result is that many carriers may choose to ignore new solutions until they absolutely require conversion to a new standard.

The Next-Generation Intelligent Network

The intelligent network is an old concept now, but the idea is still a powerful one. By embedding intelligence in the network in the form of databases and computers capable of accessing them, and by devising applications that extend the capabilities of the in-place network through the judicious use of those network-based databases, the network suddenly begins to generate revenue in new ways. Value-added services are designed to eke more out of the legacy network, and they do: Consider the revenues generated by Caller ID services alone. These information services add revenues to a revenue stream (voice) that is essentially flat. Next-generation services are those that will become available as the network evolves through the addition of optical bandwidth, more effective optical management, and more evolved routing capabilities (MPLS and GMPLS).

Although MPLS was not specifically designed for the metro market, it is carving out a significant niche for itself there. As we discussed earlier, its primary function is to enable routers and switches to designate end-to-end paths through a packet or cell-based network with specific QoS characteristics implied. As a result, carriers can support connection-oriented services such as enterprise VPNs and ATM, a variety of traffic engineering requirements, and bandwidth management on a network-wide basis. Because it is compatible with both ATM frames and IP packets, MPLS supports the evolution of IP switching, the well-known hybrid technology that incorporates the best of both IP and ATM.

MPLS was originally targeted at packet networks with the specific goal of providing a migration path that would ultimately eliminate the need for ATM, the sole source of QoS in modern packet networks. Generalized MPLS, however, is designed for all-optical networks and works with SONET, SDH, and DWDM.

MPLS and GMPLS may well serve as the integration layer for metro networks by providing intelligent management and QoS guarantees for DWDM-enhanced optical networks that transport IP packets. It could offer end-to-end bandwidth provisioning, traffic engineering, network or lambda management, and monitoring/redundancy services.

Network Management

The last thing implemented and the first thing missed, network management is one of the most important aspects of competitive networks. Without management, the network cannot be monitored, provisioned, maintained, or able to generate invoices for the usage of resources. It is perhaps the single most important differentiator of all, because it permits network administrators to define the capabilities of their network and to develop service offerings based on that definition.

These efforts are designed to improve network efficiency, manage the increasingly unpredictable (and massive) packet volume that networks are required to transport, reduce the overall cost of providing network services, and reduce provisioning intervals (a significant competitive advantage). In short, the goal is to (1) reduce costs, (2) increase revenues, or (3) both.

Consider, for example, the evolution of SONET and SDH. It is a massively overhead-heavy protocol model that manages a wide array of functions. At one time, those functions were necessary because no other network protocol was capable of performing them. Today, however, many of those functions are considered to be redundant or no longer required, given the changes that have occurred with the characteristics of transported data.

Today SONET and SDH are being positioned to perform a reduced subset of tasks. These include 50-millisecond protection switching to a backup span, remote element management to ensure that any device on the ring can be surveilled and managed remotely, enhanced performance monitoring, end-to-end path management, and the full complement of *Operations, Administration, Maintenance, and Provisioning* (OAM&P) capabilities that the technology has come to be known for. All these exist in traditional legacy SONET/SDH implementations, but many other capabilities that have now been deemed largely unnecessary also exist, so the intent is to "prune the capability tree," as it were. The bottom line is this: Networks by and large are migrating to a packet model, and SONET and SDH were never designed to handle packets. The intent is to evolve the legacy infrastructure so that it lends its considerable capabilities to the IP world while at the same time preserving the investment that service providers have made. Given the interest in evolving multiservice network infrastructures, it only makes sense that legacy technologies, to the greatest degree possible, will be preserved with a certain degree of modification.

SONET and SDH, which will be discussed in detail later in the book for those readers interested in learning more about the details of the technology, are really expanded and more capable versions of T- or E-Carriers—T1 on steroids, if you will. They provide a channelized or unchannelized "pipe" capable of carrying a wide variety of payload types, provided those payload components fit within the parameters of the framing structure. The new SONET or SDH is a multiservice-friendly technology that is more adept at transporting packets, a characteristic that is particularly important in metro networks. The only downside today is that because this evolved architecture is still relatively new, it is largely proprietary. However, the ITU-T has created Draft Standard G.7041, which defines a *generic framing procedure* (GFP) for the evolved SONET/SDH infrastructure. The bulk of the standard was written in 2001, and while work on it continues, it is expected to be complete by the end of 2002 or the beginning of 2003. In effect, G.7041 removes components of SONET and SDH that are deemed to be unnecessary for modern networks, leaving what many call a "thin physical layer" deployment that preserves required capabilities while eliminating those that are not. It will be implemented in two forms: Frame-Mapped (GFP-F) and Transparent mode (GFP-X).

Frame-Mapped mode will be used for the encoding and transport of protocols that have already been framed or packetized by the originating client, such as the *point-to-point protocol* (PPP), IP, Ethernet *Media Access Control* (MAC) frames, or even Frame Relay frames. Transparent mode is designed to address the growing requirements of LAN and storage network installations and is designed to transport block-mode traffic such as Gigabit and 10-Gigabit Ethernet, *Fibre Channel Connectivity* (FICON), *Enterprise System Connectivity* (ESCON), Fibre Channel, and other *Storage Area Network* (SAN) protocols.

Standards developers cite numerous advantages to the generic framing structure approach, including improved bandwidth efficiency due to multiplexed packet-based data streams, the capability to deliver granular QoS, significantly lower overhead than prior technology solutions, backward compatibility with existing implementations, a multiservice-friendly approach, a standards-based mapping scheme for multiple protocols, and, perhaps most important, a design that is friendly to current, transitional, and future network and service architectures.

The generic framing structure concept is not necessarily a universally acceptable solution, however. Because it is built around the SONET/SDH model, it will be most attractive to companies with a large installed base of legacy optical technology—the incumbent carriers. It is also (to use a term from *The Hitchhiker's Guide to the Galaxy*) inextricably *intertwingled* with other emerging technology standards such as optical burst switching and resilient packet ring, which means that its adoption and commercial success may be a bit slower than many would like.

An Aside: Resilient Packet Ring (RPR)

RPR has been standardized under IEEE 802.17 and is designed to interoperate with any packet- or frame-based protocol, not just Ethernet. It is designed to support the delivery of so-called carrier-class packet services through the adoption and use of expanded redundancy, traffic engineering, and deterministic resource management.

Expanded redundancy refers to the fact that RPR relies on traditional ring survivability techniques, including a 50-millisecond switchover from a primary to a backup ring, node awareness, and ring wrapping. *Traffic engineering* refers to the fact that RPR is designed to allocate bandwidth on an as-needed basis, using a network-wide (or ring-wide) bandwidth allocation technique. Finally, *deterministic resource management* speaks to the fact that RPR replaces traditional ring-buffering techniques with ingress node queuing and packet discards, a far more intelligent packet management technique. These will be discussed in more detail later in the book.

Anatomy of the Metro Network: The Edge

Metro networks reside at the edge of the WAN or core backbone network and touch both the user and the core. As a result, they incorporate both edge and core devices. Edge devices operate at the frontier of the network and are responsible for protocol internetworking, traffic concentration, QoS policy enforcement, and traffic flow discrimination. They provide the access capabilities of the metro network and serve as the interface to the wide area transport domain. This is also the place where IP is often first encountered as a network-layer protocol used to implement multiprotocol or multiservice networking.

The Multiservice Metro Network

Multiservice networking defines the capability of a single network to transport a variety of traffic types over the same physical infrastructure while preserving their various requirements for transport quality. In today's world, six primary service types are targeted for this type of network:

- *Voice,* which includes not only so-called carrier-class voice, but also IP telephony and other alternatives.

- *IP,* which includes Internet access, long-haul IP transport (with or without QoS), and interworking among routers.

- *Private line,* which includes both channelized and unchannelized T1 and E1, point-to-point Ethernet transport, DS3, E3, and SONET/SDH facilities.

- *ATM,* including QoS characteristics such as channel concatenation, service adaptation and emulation, and QoS-aware transport support for DSL access, cable access, and voice, video, and data services.

- *LAN,* including VPNs, LAN extensions, *virtual LANs* (VLANs), *transparent LAN* (TLAN) services, and LAN-based edge switching.

- *Wavelength-based services,* which rely on DWDM to offer scalable and configurable bandwidth on demand.

It should be clear that because of their unique location in the overall topology, metro networks must provide a collection of key functions that distinguish them from the edge and core. These include

- *Demarcation*, the logical separation point between the core and the edge, or some other designated network regions

- *Adaptation*, the technique used to match a transported payload to a transport medium

- *Aggregation*, the process of combining multiple traffic flows for simultaneous transport across a shared path

- *Transport*, or transmission

- *Edge switching,* the process of establishing the initial path at the edge of the network before being handed off to the long-haul portion of the network

- *Core switching,* long-haul, high-speed transport

- *Grooming,* the addition or deletion of certain designated channels at an intermediate point along the way

- *OAM&P*, the standards functions performed in the management of a telecommunications network
- *Billing and service delivery,* as defined by the customer

Furthermore, because they are visible to the customer (because they *touch* the customer) and also connect to the core transport network, metro networks must be survivable, flexible, reliable, cost effective, easily configurable, and manageable. They must also offer a broad array of capabilities, including differentiable QoS, service transparency (transport and payload independence), broadcast, and point-to-multipoint connectivity.

Much of the energy that has propelled the service providers toward a multiservice philosophy and which in turn has caused the system manufacturers and therefore the component manufacturers to turn their attention in the same direction comes from the ongoing technology convergence phenomenon. Technology convergence defines the desire of service providers to simplify their business operations by reducing the complexity of the network through reductions in hardware, software, and protocols that they must support.

The common misconception is that service providers are working hard to migrate traffic to a single, all-IP network, eliminating the need for separate support, operations, design, provisioning, repair, and sales organizations for each of their networks: ATM, Frame Relay, IP, wireless, the *Public Switched Telephone Network* (PSTN), X.25, and others.

This is indeed a misconception, although not entirely. Service providers *do* want to simplify their networks by reducing the number of elements they must manage, the amount of enabling software they must support, and the number of protocols they must carry. However, they are also quick to say that their intent is not to retire legacy technologies such as T1/E1, SONET/SDH, and others, because these technologies do, after all, generate the bulk of service provider revenues. Instead, they are making seminal decisions about the overall architecture of the network, taking into account the fact that the core and the edge are distinctly different in both form, function, and the developing models that enable legacy and new technologies to exist side by side.

The device that makes this change in the network possible is the multiprotocol (or multiservice) *add-drop multiplexer* (ADM). The concept is simple. A high-speed backplane in the device accepts input from an array of *input/output* (I/O) cards, each of which is designed to process a different protocol: ATM, *Integrated Services Digital Network* (ISDN), IP, PPP, Frame Relay, voice, and a host of others. QoS software enables traffic prioritization

Figure 1-13
A multiservice ADM

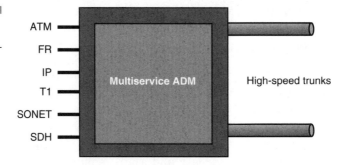

and queue management; the result is network simplification. A schematic representation of a multiservice ADM is shown in Figure 1-13.

Network Inversion

The network is undergoing an inversion of sorts in which the distinct functions of the edge and core regions of the network are becoming (for the first time, really) clear and distinct.

Historically, the core of the network has had a variety of functions. It provided high-speed, wide-area transport, the signaling intelligence necessary to carry out efficient network routing, many of the applications that provided functionality to customers, and rudimentary protocol discrimination. In the last few years, however, that model has become untenable because of customer application diversification and the growing demand for increasingly granular service differentiation at the edge of the network. As a result, many of the functions that have traditionally resided within the network core are migrating to the edge. For example, protocol discrimination (multiprotocol capability) and QoS-based policy enforcement, formerly done in the core switch or router, are now performed at the edge of the network by intelligent edge devices that aggregate traffic, prioritize it according to a rule set established by the customer, and communicate with core transport resources to ensure that each logical traffic stream is handled according to its QoS requirements.

Furthermore, bandwidth has begun to migrate from the core to the edge as the network inversion continues. Recent regulatory decisions that make the universal provisioning of broadband access infinitely more appealing to the incumbent service providers will cause them to begin the task of pumping additional capital into the edge of the network in short order. High-speed access that is widely available will spark the creation of broadband-dependent applications, most notably those dependent upon multimedia content. The result of this cascade of functionality is the creation of the *multiservice edge.*

Consider the facts: Service providers have spent hundreds of millions, if not billions, on their access and transport infrastructures, doing such a good job that they have now been rewarded with the gift of commodity. Networks are capital-intensive resources, yet the revenues they yield for access and transport are in rapid decline—some would say free fall. They need help, and they need it now.

The manufacturers are faced with a similar conundrum. Advances in semiconductor technology allow them to create entire systems on a chip that have the capability to process multiple disparate protocols. This almost seems too good to be true for both the embattled incumbent providers and the manufacturers. Here is a revenue formula that works! By building multiservice networks using lower-cost technology, hybrid solutions that serve a wide array of clients can be created. However, it's a bit more complicated than that.

In the first place, incumbent service providers lack long-term experience with multiservice, multiprotocol networks. Second, many of the companies that are designing and manufacturing the multiservice edge devices are new, unknown, not yet standardized, and lacking in one critical element: the trust of the marketplace. Finally, if the rollout of such an innovative and even radical new service offering is not handled well, it can further confuse the market, generating greater levels of distrust and discouraging investment.

Perhaps an example is in order. Ethernet's ascendancy to the role of heir apparent to the burgeoning metro broadband market is a great thing technologically, as is the early perceived success of such innovations as resilient packet ring and optical burst switching. However, as good as these solutions may be, they confuse the market as players ask questions about the relative viability of Ethernet versus legacy transport, SONET and SDH versus RPR, and the apparent functional universality of IP in the LAN, MAN, and WAN.

Cisco's Central Role

In 1998, Cisco Systems acquired Cerent, a manufacturer of ADMs targeted at the optical network. Cerent's charter, described in a press release issued at the time of the acquisition, was "to provide next generation SONET ADM equipment that is a fundamental building block in voice and data networks and used to add and remove lower speed traffic from higher speed optical rings. Service providers will use Cerent's product in the access portion of networks." These boxes provided transport for both voice and data services, one of the first realized examples of technology convergence, and a compelling reason for Cisco to acquire them. Companies in the industry began to refer to these boxes as *multiservice access platforms*, a name that soon gave way to *multiservice access devices*. Because of their multifunctional architecture, these systems are less costly, less power-dependent, and have a smaller footprint than their predecessors—critical issues for competitive providers that must work within the confines of a collocation environment. They are designed to seamlessly offer both transport functions and higher-level service. They can do this because of the capabilities of new chipsets that incorporate multiple layers of functionality.

Although ideal for the aggregation and transport of circuit-based traffic (voice, some data), traditional SONET/SDH ADMs do not lend themselves to the efficient transport of bursty packet data such as that generated by IP or LAN environments. By aggregating traffic at the edge from multiple sources, networks can be made significantly more efficient and cost effective. Consider the typical access network. Studies indicate that even in the enterprise data space, the local loop is woefully underutilized. As a full-time dedicated resource, this is a very expensive network component for both the customer and the service provider. If technologies exist to drive traffic into that loop to statistically increase its utilization, so much the better.

Other factors are equally important. As functionality has moved from the core to the edge, certain key capabilities have benefited greatly from the shift. Some of these are described here.

Route Management

The arrival of route control protocols like MPLS permits optimal route selection to be performed at the edge of the network instead of relegating the responsibility to core transport routers. This results in far more efficient network utilization but also yields a route choice—which in turn leads to redundancy and route diversity.

Security

Because enterprise firewall routers are typically standalone devices and by design are found between the private enterprise network and the public transport network, it only makes sense that the security function would naturally migrate from the core of the network to the edge and become the responsibility of the end user.

Data Archiving and Caching

The growth and interest in SAN has driven interest in data access to the edge and beyond as ASP architectures have emerged that place the data stores beyond the realm of the network core. Additionally, IT planners have come to realize that if caching stores are designed into the network architecture that enable frequently accessed database records to be cached locally, the impact on the transport network can be reduced substantially. Instead of retrieving data from a remote location, the data is archived locally and retrieved from a cache server.

Policy Enforcement

When policies are enforced from a centralized location (a network core), all traffic tends to be treated equally. When queuing and delay policies are established locally (at the edge), however, they tend to be customized to local traffic behavior patterns. Thus, the decision to enforce policy at the edge rather than at the core tends to make for a more efficient network.

VPNs

Anytime a public network can be used to deliver secure private service, it should be done. VPNs enable a service provider to, in effect, "resell" the same physical network resource over and over again. VPNs are possible because of the capability of edge devices to "tunnel" through the public network infrastructure, creating a virtual dedicated channel for customer-to-customer traffic.

Of course, the multiservice concept does have disadvantages. First of all, the devices tend to be complex, which often leads to expense. Second, although the complexity of these boxes makes it possible for them to offer every service at every network location, every service isn't *needed* at every network location. As a result, the service provider often pays for a capability that they do not require. Overall, however, the benefits of a multipurpose edge device far outweigh the liabilities of complexity and cost. As metro continues to evolve and grows to become a major market segment, the multiservice capability will become central to the success of all levels of the networking food chain, particularly for the service provider.

Let's stop for a moment and restate the advantage of this architecture. The service provider aggregates traffic via a single connection to a multiservice edge device. The customer enjoys low cost, low complexity, and a high degree of connectivity. In short, everyone wins.

The Metro Network: Access

Access refers to the ability to connect to the resources offered by a metro network. As Figure 1-14 illustrates, any number of access options are available, including traditional dial-up, ISDN, DSL, cable modems, various forms of Ethernet, wireless, Frame Relay, and traditional dedicated private line such as T1 and E1. Keep in mind that access occurs at the edge of the network and that the edge is responsible for a number of functions, including traffic discrimination, concentration, QoS determination and enforcement, and protocol interworking. The edge is also the place where IP is first found in the network, because it is at the edge where other protocols are packaged for transport across wide area IP networks, the growing standard for network-to-network interconnection.

Metro networks have only recently become a "standalone domain" within the hierarchy of network architectures. This "coming of age" of the metro space is a result of the natural evolution of both networks and the corporations that use them. During the past 20 years, corporate structures have evolved from centrally managed (and clustered) models to more distributed models. They have come to realize that to be competitive they must be physically located as close to their customers as they can possibly be, a philosophy that conflicts with the concept of a centralized business operation. Furthermore, they have found themselves affected by quality of life and

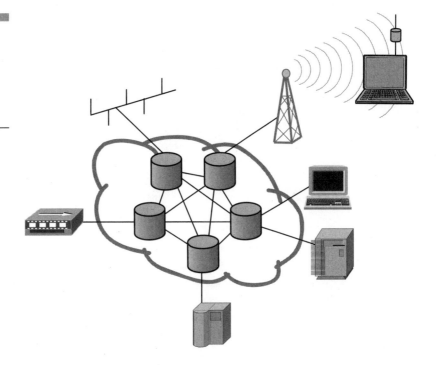

Figure 1-14
Access comes in many flavors, including wireless, DSL, ISDN, cable, and a variety of others.

environmental concerns such as the demand from employees to telecommute from home and Federal requirements such as the United States' 1990 Clean Air Act. The Act attempts to reduce airborne pollutants, the corporate response to which has been to allow workers to work from home, thus reducing vehicle emissions overall.

The result of this is that networks have had to evolve to ensure that employees of distributed corporations enjoy connectivity characteristics that are of the same quality they enjoyed when working at their single business location. The more or less simultaneous development of SONET and SDH, cost-effective ring architectures, and bandwidth management applications, combined with widely available and cost-effective high-speed access technologies, means that metro networking is effective and that it has helped to redesign corporate interconnection and communication models.

Eight access technologies are the most successful in metro applications: 56 Kbps dial-up modem, ISDN, DSL, cable, Ethernet, wireless, Frame Relay, and private line.

56 Kbps Dial-Up

Even though it is viewed as a true legacy solution, *56K dial-up access* continues to be widely used, even by remote corporate employees. DSL, ISDN, cable, and wireless are not always available, so 56K continues to enjoy widespread use. The recent release of the V.92 modem standard, which improves on its V.90 predecessor by offering new services such as Quick Connect, Modem-on-Hold, and PCM Upstream.[2]

ISDN

ISDN, sometimes referred to as "the technology that took 15 years to become an overnight success," is now enjoying something of a renaissance because of the failure of some of its technology peers to perform. When DSL first arrived on the scene, it was viewed as the solution that would finally drive a stake into the heart of ISDN. However, because of deployment issues, performance problems, and availability limitations,[3] DSL has been slow to succeed in many markets. As a result, ISDN has enjoyed a much higher uptake rate in the past 24 months. Even though it is limited to 128 Kbps (by bonding the two B-Channels into a single channel), it works, is internationally standardized, and is widely available at a (for the most part) reasonable price. Current compression technologies make 128 Kbps a reasonable solution for moderately bandwidth-intensive applications, including video.

DSL

DSL, when available, is a very good access solution. With downstream bandwidth ranging from 128 Kbps to as high as 52 Mbps (depending on the version chosen), it offers a variety of access solutions for even the most

---

[2]Quick Connect reduces the typical connect time by 50 percent or more by "memorizing" the line characteristics to reduce the length of successive handshake procedures. Modem-on-Hold enables a user to receive an incoming call while maintaining a previously established connection to the Internet. PCM Upstream increases the upstream data rate from 33.6 Kbps to a maximum of 48 Kbps, a 30 percent increase.

[3]Bellsouth is the exception: In January 2002, they announced that the number of working installs of DSL (more than 620,000 in territory) had far exceeded their and the market's expectations.

bandwidth-hungry applications. Although it has enjoyed its greatest success in the residence marketplace, it is also widely used for corporate interconnections. For example, ISPs often use DSL as the backbone connector between their router pools and the *Incumbent Local Exchange Carrier* (ILEC) access network. This is a technology that will continue to grow in importance; watch it closely.

Cable

Cable is a bit out of place in the context of enterprise access, but because it represents a viable broadband access solution for the *small office, home office* (SOHO) and telecommuter, it would be irresponsible to ignore it as an option. Cable offers asymmetric access, with a far higher downstream channel (roughly 1 Mbps) than its upstream counterpart. This design stems from historical usage; most cable modem users download large files from web servers using relatively small upstream commands that yield massive amounts of graphics-heavy content downstream. As home workers become more common, however, this model may begin to show its limitations as their requirements to upload large files grow. Remember also that cable is a shared downstream medium. The more users that share the channel, the less bandwidth is statistically available for each user. Thus, it suffers from its own success in the market.

Ethernet

In the enterprise access domain, *Ethernet* has gained a degree of notoriety among its various access technology peers by effortlessly morphing into multiple forms as the demand for different solutions has evolved. Originally conceived as a 1 Mbps LAN data-link-layer protocol, Ethernet has expanded beyond both the LAN world and its multimegabit limitations.[4] Such companies as Telseon, Yipes, Atrica, and others are now deploying Ethernet as a transport solution largely because of its popularity, simplicity, availability, and bandwidth. Ethernet is now available in 10 Mbps, 100 Mbps

---

[4]A word about nomenclature: Ethernet is actually the name coined by Bob Metcalfe in 1976 for his concept of a distributed packet network architecture—for all intents and purposes a product name. The technology (protocol) that defines Ethernet is called *Carrier Sense, Multiple Access with Collision Detection* (CSMA/CD).

(Fast Ethernet), 1 Gbps (Gigabit Ethernet), and 10 Gbps versions. When deployed as a point-to-point transport option, *Carrier Sense Multiple Access with Collision Detection* (CSMA/CD) is not used, because no collisions must be managed. CSMA/CD manages the shared medium by detecting and correcting collisions that occur when two stations transmit simultaneously.

Wireless

Wireless access is, in fact, another option that has resulted from the limitations of wired alternatives. The two most successful options today are the *Local Multipoint Distribution Service* (LMDS) and the *Multichannel, Multipoint Distribution Service* (MMDS). LMDS operates in the 28 to 31 GHz band, while MMDS operates in the 2.5 to 2.7 GHz range. LMDS offers 155 Mbps over distances of up to 10 miles, while MMDS offers a much lower 10 Mbps, but over a distance of up to 35 miles. Equally important has been the success of 802.11b, the 11 Mbps wireless Ethernet standard that also goes by the name of Wi-Fi (for *Wireless Fi*ber or Wireless Fidelity.)

Frame Relay

Frame Relay seems to have dropped off the face of the earth in many cases; it just isn't discussed much anymore. Nevertheless, it continues to sell well as a switched alternative to dedicated private line. The primary advantage it offers over private line is its pricing algorithm; no distance component is included in the scheme, which means that Frame Relay offers distance-insensitive pricing. Under normal circumstances, this would not be an issue in metro environments, since most distances are relatively local. The breadth of metro areas, however, is beginning to blur a bit. In the Los Angeles market, for example, it is not uncommon for a metro network to require runs of as far as 80 miles, a distance that could result in a prohibitively expensive circuit when price is based on circuit miles. Frame Relay also offers access speeds ranging from 56 Kbps to 45 Mbps.

Private Line

Of course, a market for *dedicated private line* still remains. The price of a T1 or E1 has plummeted in recent years; a circuit that cost $45,000 per year five years ago costs $4,000 today. The result is that the popularity of dedicated facilities, particularly given renewed concerns about security, has gone up as the price for such circuits has gone down.

Metro access options will continue to diversify in response to growing requirements for bandwidth and QoS guarantees from an increasingly demanding customer base. The metro network space itself will continue to evolve in lockstep with the corporations that make use of its capabilities and will be called upon to carry a growing diversity of application content to the transport network.

Anatomy of the Metro Network: The Core

Core devices, on the other hand, operate within the transport region of the metro network and are responsible for carrying out the service requests that originate within the edge domain—specifically the high-speed transport of data packets. Devices found here include large IP routers, ATM switches, and hybrid platforms that include MPLS and other newer protocol suites.

Within the metro environment, *transport* refers to the process of moving traffic from one node in the metro network to another, but always within the metro space. For example, a corporation might have multiple business locations within a metropolitan area, all of which must be interconnected to provide intra-enterprise transport that is seamless, application independent, and completely transparent. The characteristics of such a multipurpose transport network are "unlimited" bandwidth, rapid provisioning of bandwidth to users, and QoS awareness. QoS awareness simply means that the network is aware of the fact that different applications place varying demands on network resources and is able to adjust to and satisfy those demands in real-time or near real-time.

Early metro transport networks relied on legacy copper infrastructure for the most part, although some microwave was also used for building-to-building interconnections. As bandwidth demand grew and the diversity of bandwidth-hungry applications and numbers of users climbed, copper began to show its limitations as a transport option. It is still used for some applications ("If it ain't broke, don't fix it"), such as simple multichannel voice installations for *private branch exchange* (PBX) networks, but for data and multimedia transport it no longer suffices.

Beginning in the mid-1980s metropolitan fiber networks began to appear, thanks largely to the efforts of such companies as *Metropolitan Fiber Systems* (MFS, now part of WorldCom) and *Teleport Communications Group* (TCG, now part of AT&T). These companies came into existence with the sole intention of competing with the incumbent providers by building dual-entry optical rings throughout metropolitan areas and selling services on the value of their support and survivability. The model they established is the most widely accepted model today and the basis for most carriers' metro networks. The fact that their networks were *dual-homed*, meaning that all served buildings were connected to both the primary and secondary ring, which in turn entered the buildings from two separate sides of the building (route diversity), meant that service interruptions were almost nonexistent.

This ring architecture is primarily SONET- or SDH-based today, although other architectures have recently emerged. SONET and SDH networks typically comprise two rings that transmit in opposite directions.

In the event of a failure of the primary ring, a switchover occurs to the secondary ring, which then carries all live traffic. The switchover takes place within 50 milliseconds and is transparent to the user. If both rings are breached, the ring reconfigures into a toroidal shape, losing its capability to survive another fiber cut but preserving the capability to transport traffic. Ring topologies will be discussed in greater detail later in the book.

SONET and SDH have found themselves under fire of late as some network professionals have tried to relegate them to the boneyard of legacy technologies as newer technologies have arisen. One commonly heard complaint is the perception that SONET and SDH are overhead heavy, particularly when operating in concatenated mode. True enough, but that overhead is highly functional and so far nothing new has come along that offers such a complete package with less overhead required to do it. New technologies have arrived, however, and they are enjoying some degree of success. One of these comes to the table from Cisco and is called the *Spatial Reuse Protocol* (SRP).

SRP enables devices such as routers to be interconnected by dual fiber rings using SONET and/or DWDM. The rings offer redundancy and operate at 10 Gbps with a ring-to-ring switchover speed of less than 50 milliseconds. In MANs, SRP dramatically lowers the cost of network deployment for corporations or ISPs looking to interconnect multiple sites by eliminating the need for an expensive ATM switching infrastructure. The one thing SRP does not adequately offer today, however, is a well-defined strategy for QoS, something ATM does extremely well.

Nevertheless, SRP has found a home in the optical world because of its use in RPR, mentioned earlier. RPR is a network transport scheme for fiber rings. The IEEE began the 802.17 RPR standards project in December 2000 with the intention of creating a new MAC layer that offers the same protection capabilities that SONET and SDH make possible, without the overhead. Fiber rings are widely used in both MAN and WANs; these topologies, unfortunately, are dependent on protocols that are not scalable for the demands of packet-switched networks such as IP. The 802.17 RPR working group promotes RPR as a technology for device connectivity. Issues currently being addressed are bandwidth allocation and throughput, deployment speed, and equipment and operational costs. Both RPR and SRP will be discussed in greater detail later in the book.

Optical Burst Switching

A related technology is *optical burst switching* (OBS). OBS supports real-time multimedia traffic across the Internet, a natural pairing since the Internet itself is bursty. OBS is particularly effective for high-demand applications that require a high bit rate, low latency, and short duration transmission characteristics.

OBS uses a "one-way reservation" technique so that a burst of user data, such as a cluster of IP packets, can be sent without having to establish a dedicated path prior to transmission. A control packet is sent first to reserve the wavelength, followed by the traffic burst. As a result, OBS avoids the protracted end-to-end setup delay and also improves the utilization of optical channels for variable bit rate services. By combining the advantages of optical circuit switching and those of optical packet or cell switching, OBS facilitates the establishment of a flexible, efficient, and bandwidth-rich infrastructure.[5]

Another transport technology that has come booming into the forefront of technologists' minds in the last year is Ethernet. Although positioned as an alternative to SONET and SDH, it is important to remember that

---

[5]For more information about RPR, please visit the 802.17 Alliance Web site at http://grouper. ieee.org/groups/802/17/. For information about OBS, please see the article posted at the Light Reading Web site at www.lightreading.com/document.asp?doc_id=9016&page _number=8.

SONET and SDH are physical-layer (layer one) protocols, while Ethernet (CSMA/CD) encompasses layer two as well. SONET and SDH are transmission technologies; Ethernet is a framing and switching technology.[6] Ethernet can be deployed as a transport technology and can certainly be transmitted over optical fiber, even carried over SONET and SDH. What it brings to the transport world is simplicity, low cost, scalability, and broad acceptance and understanding. What it lacks, and what SONET and SDH offer, is protection, resulting in QoS preservation. This is the reason that 802.17 and the not-yet-mentioned X.86 standards are so important. 802.17 provides the means to deliver the robustness and survivability typically available only in SONET/SDH rings via Ethernet, while X.86 offers a standard scheme to map Ethernet frames into SONET/SDH transmission streams.

So what does this mean? It means that SONET and SDH are not going away any time soon: Too large an embedded base has been established for that to happen. It does, however, mean that for the near term, we will continue to see side-by-side coexistence of Ethernet with SONET and SDH, particularly in the metro network domain.

Freespace Optics

Another technology that should be watched is *freespace optics*. Freespace optical solutions rely on open-air laser transmission between buildings to create low-cost metropolitan networks. Most industry estimates today claim that more than 80 percent of all business buildings are *not* served by fiber. Freespace optics, therefore, would seem to be a good alternative solution and can be used as mesh alternatives.

Several companies have entered the freespace game, including AirFiber with its 622 Mbps OptiMesh™ product and Terabeam, which provides wireless optical-like connectivity at 5, 10, and 100 Mbps. Both companies offer high degrees of survivability and reliability.

---

[6]We should note here that although Ethernet is used today as a transport technology, CSMA/CD is not. When Ethernet is deployed for transport, only its simple data-framing structure is used: Neither collision management nor the other media management capabilities of the protocol are used.

Asynchronous Transfer Mode (ATM)

Although many criticize ATM as an overhead-heavy legacy solution, it continues to be installed at a steady pace around the world because of its capability to assign and maintain guaranteed levels of service quality at all levels of the network—local, metro, and wide area. It is a comparatively expensive solution, but when customer service quality is at risk, cost becomes less important. Commonly seen in wide area installations, ATM has begun to show up in the metro environment as well. Its price point is dropping, making it more palatable to network designers. IP is also popular because of its universality, but typically only when transported by another protocol capable of preserving QoS.

Metro Topologies

Within the metro domain, three network topologies or architectures have emerged that are vying for a place at the technology table: point-to-point, ring, and mesh. Each of these have advantages and disadvantages, including design challenges, installation issues, the capability to monitor and survey the overall network (or not), and concerns over engineering complexity and design. Because all three of these topologies will be evident in our studies of the metro network throughout the book, we will discuss them briefly here.

Point-to-Point Architectures

Point-to-point physical networks are cost effective under certain circumstances such as short-reach installations (as many metro network segments are) and where the single span of the circuit can be enhanced through the use of DWDM as a way to provide bandwidth multiplication. In many instances, companies already have dedicated facilities in place; as long as they provide adequate bandwidth and offer redundancy through carefully laid-out network plans, they will continue to suffice.

Ring Architectures

As we mentioned earlier, rings represent the current topological standard for metro networks, although that preference is shifting in favor of alternative approaches. Rings have been in widespread use since the mid-1980s when SONET and SDH first made their way onto the networking scene.

Today, rings are commonly used because of the enhanced survivability that they provide. Ring problems can be caused by electronic failures, digital cross-connect and multiplexer database failures, internal power failures, aerial cable failures, and, of course, the infamous "backhoe fade" that results from an underground cable cut. The first three are difficult to protect against, although hardware designers typically build in component redundancy to reduce the chances of a device-related failure. More than 75 percent of all optical network failures, however, result from fiber breakage, either aerial or underground. As a result, anything that can be done architecturally to protect against physical network disruption will go a long way toward ensuring the survival of the overall network. The most common solution is to deploy redundant paths in the most critical spans of the network, and this is most commonly done using ring topologies.

The Unidirectional Path-Switched Ring (UPSR)

The *unidirectional path-switched ring* (UPSR) is comprised of two fiber rings interconnecting the add-drop multiplexers that make up the network connectivity points. One fiber serves as the active route, while the other, which transmits in the opposite direction, serves as a backup. In the event of a fiber failure between two adjacent nodes, the SONET or SDH overhead bytes convey the failure information from device to device around the ring, one result of which is that the traffic is switched to the backup span within 50 milliseconds. The backup span then carries the traffic, this time in the opposite direction, ensuring the integrity of the ring.

Unfortunately, this architecture results in a problem. Consider the ring shown in Figure 1-15. Under normal operating conditions, traffic flows from node one to node two to node three and so on. Traffic going from node two to node three only has to traverse a single hop. The other direction, however, is a very different story. Traffic from node three to node two must traverse a four-hop route, resulting in an asymmetric two-way transmission path. For geographically small rings, this is not a problem. For larger rings, how-

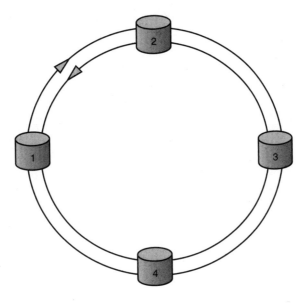

Figure 1-15
Normal traffic flows
in an optical ring

ever, it can be a significant problem because of synchronization and latency concerns. For example, an application, expecting symmetric transmission and reception, might adjust its timers for equal arrival and departure rates. If it does this, however, it will see data constantly arrive early, while the other end will see it arrive late—or vice versa. Clearly, a better solution might be in order.

Bidirectional Switched Rings

Four-fiber, bidirectional switched rings eliminate the asymmetric problems that characterize unidirectional two-fiber rings. These bidirectional rings can be engineered in two ways: as span-switched rings or line-switched rings. Both provide high degrees of protection for data traveling on the ring, and both send traffic bidirectionally to ensure symmetrical delay in both transmission directions.

Span-switched rings recover from span failures by switching the bidirectional traffic on the active pair of fibers to the backup pair, but only between the nodes where the failure occurred. This is clearly a better overall design than the two-fiber UPSR, although it does have one common failing: All too often the active and backup fiber pairs are found in the same conduit, which means that a truly diligent backhoe driver *will* take down the ring by cutting all four fibers.

An alternative that offers better survivability is the *bidirectional line-switched ring* (BLSR). These are complex architectures that require significant bandwidth management and control. In a BLSR, half the traffic is sent on the clockwise ring, while the other half is sent on the counterclockwise ring. This results in a much more efficient use of bandwidth and significantly improved performance. If all four fibers are cut between two adjacent nodes, both the active and the protect paths wrap to preserve the integrity of the ring and the data that it carries. Thus, the BLSR is the most popular configuration in the SONET world, although UPSRs are still more common outside of North America. In fact, *four-fiber BLSRs* (4F BLSRs) are often seen deployed for long-haul applications and are usually diversely routed as additional protection against physical failure. Even in the United States, where four-fiber rings have been the norm for some time, two-fiber rings (2F BLSRs) are becoming more common. Table 1-1 outlines the characteristics of the three different ring types.

The Two-Fiber BLSR Option

Within metropolitan areas, the 2F BLSR has become a common deployment alternative. It guarantees survivability by "gerrymandering" the available bandwidth. Signals are transmitted on both fibers simultaneously. Each multiplexer on the ring is configured to seek its primary signal from one ring or the other, and each span between multiplexers carries the config-

Table 1-1

UPSR, 2F BLSR, and 4F BLSR

Characteristic	UPSR	2F BLSR	4F BLSR
Delay	Asymmetric	Symmetric	Symmetric
Deployment	Metro	Metro	Long-Haul
Multiple failure protection	No	No	Yes
Initial cost	Medium	Medium	High
Augmentation cost	Low	Medium	Low
Complexity	Low	High	Medium
Efficiency	Medium	Medium	High

ured capacity for both active and protect traffic to ensure bandwidth availability in the event of a catastrophic failure of the ring. This particular model has become more and more common.

Mesh Architectures

Mesh networks are defined as networks in which every node on the network is connected to every other node by a collection of point-to-point links. These networks can be partially meshed, as shown in Figure 1-16, or fully meshed, as shown in Figure 1-17. The choice of partial or full-mesh connectivity is a function of desired survivability levels, the nature of the data to be transported, and cost. Full-mesh networks are expensive to deploy but yield the greatest amount of redundancy for obvious reasons. The primary advantages of mesh topologies are twofold: first, they enable all connected users access to all the available bandwidth in the network, and second, routes across a meshed network can be configured rapidly. After all, a meshed network is really nothing more than a collection of nested ring networks, as shown in Figure 1-18. As a result, they offer the advantages of

Figure 1-16
A partially meshed
network

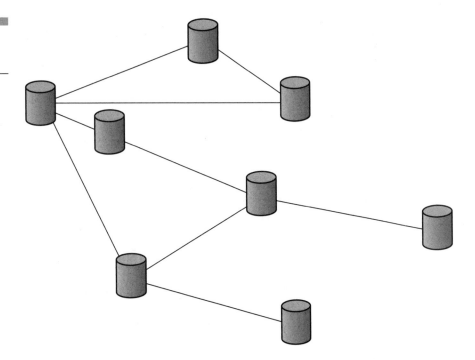

Figure 1-17
A fully meshed
network

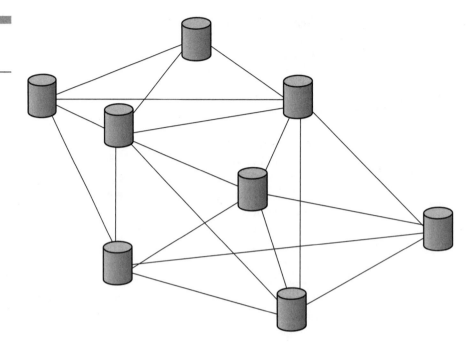

both point-to-point and ring architectures. Thus, all the desired characteristics, such as service deployment speed, redundancy, robustness, least-cost
routing, and flexibility, are available in a single topological design.

The decision to use rings versus meshes in the metro network largely
takes place in discussions about the core of the metropolitan network. After
all, the capability to have redundant routes between nodes is primarily a
metro transport issue that affects cost and service availability. Meshes are
deployed primarily in the core, while rings are found more commonly today
at the network edge, serving as traffic aggregation paths for the various
access devices (add-drop multiplexers) that connect enterprise end users to
the metro backbone. The access environment typically has less deployed
fiber than the transport core, so access rings are cost-effective solutions
since they can be shared among many different users.

It is interesting to note that a gradual migration to mesh topologies is
beginning to appear among the long-term incumbent network players. They
recognize that deployment speed and accuracy are critical differentiators in
their ever more competitive world; the capability of mesh networks to be
dynamically, rapidly, and accurately provisioned is a powerful advantage
that is not lost on the major competitors. Of course, there will continue to be

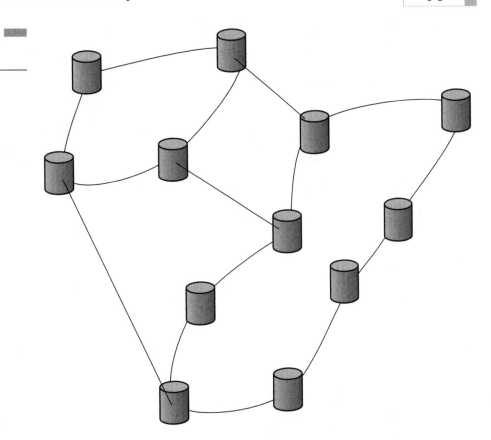

Figure 1-18
Nested rings

a place for both ring and mesh topologies. In the long term, however, meshes will most likely predominate.

Mesh topologies are typically encountered in WANs because they can easily be deployed in that environment. They are *not* typically seen in metro environments, but that, however, is changing. Because of the complex route diversity that mesh networks offer, they are beginning to be taken seriously in metro networking. They are extremely flexible and after all are really nothing more than an interlaced collection of point-to-point connections that can be managed as a series of interconnected rings. They offer higher-speed provisioning, much better use of available bandwidth (since *all* users have access to *all* bandwidth with nothing reserved for failure compensation), and no restrictions on topology or growth. They enable network managers to perform highly granular traffic engineering from any point on the network to enable the high-speed provisioning of services.

The most common mesh medium is optical fiber, although T1, E1, and other *Plesiochronous Digital Hierarchy* (PDH) technologies are also used. Traditional microwave has long been used in cities for point-to-point interconnections, although it is typically a second choice due to its susceptibility to environmental fade, its licensing requirements, and its tendency to quit working when a building sprouts up in the middle of the shot. Watch closely: Mesh metro networks will become much more widely deployed in the months ahead.

Summary

Let's go back and review the origins and characteristics of metro networks.

In the enterprise domain, a number of factors have emerged that drive the demand for metro networks. First, there has been a great deal of growth in the use of intracorporate networks, the so-called intranet model. Second, demand for bandwidth from complex applications has also grown exponentially. Consider, after all, the graph shown in Figure 1-19. Voice growth is relatively flat, while data growth is aggressive. Broadband has become a high-demand concept, including the various forms of DSL, fiber-to-the-whatever, and so on, and the demand for it is becoming as common in the metro space as it is in other network regions. Applications, too, have

Figure 1-19
Voice versus data
traffic growth

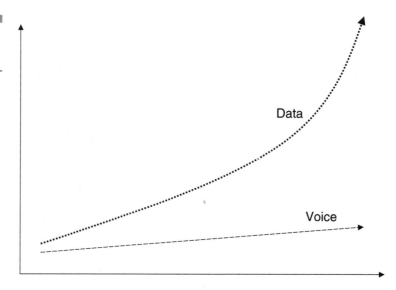

emerged that drive demand for bandwidth. These include SANs, Ethernet LAN-to-Ethernet LAN interconnections, and growing concerns about survivability, redundancy, robustness, and disaster recovery/avoidance.

The demands of the evolving marketplace include the preservation of legacy applications, route redundancy, network protection, rapid and accurate provisioning, and intelligent traffic distribution. In short, legacy access and transport technologies (Ethernet, Fibre Channel, T1/E1, and SONET/SDH) when overlaid with such demands yield the characteristics required in a metro network.

Consider Figure 1-20, which details the architecture of the traditional metro network. At the lowest layer, we find the long-haul and regional transport capabilities that typically operate at high bandwidth—nominally 10 Gbps (OC-192/STM-64), perhaps faster. Known as the long-haul optical core, this network segment is designed for the high-speed transport of massive amounts of largely undifferentiated data.

The next level up in the hierarchy is where we find the metro edge and metro core. Here we find transport options, including Fast and Gigabit Ethernet, DS3, and some SONET/SDH transport. As the name implies, this is where metro aggregation and transport occur. In this region, we find redundant optical rings interconnecting enterprise locations.

Finally, at the highest level of the network hierarchy, we find metro access. These technologies include a wide array of transport options from 64 Kbps DS0 to 155 Mbps OC-3/STM-1. Access includes passive optical network connections, SONET/SDH rings, and rings made up of ATM switches that provide aggregation services.

So what are the general characteristics of a typical metro network? First, it must be cost effective. It must be relatively inexpensive to deploy and equally inexpensive for the end user. After all, it touches the customer's

Figure 1-20

The typical metro network environment showing subtending rings

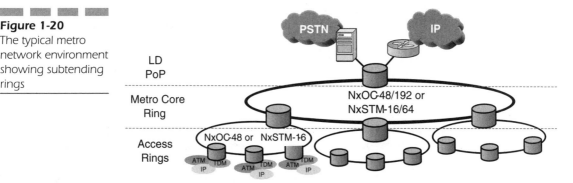

premises network, which is probably low-cost Ethernet, and the long-haul network, which in today's marketplace is equally low cost.

Next, the metro network must be designed so that adequate bandwidth is available to avoid blocking and delay problems. Some blocking and delay are normal, anticipated, and, contrary to popular belief, required to serve as cost control (metering) parameters. After all, if bandwidth is perceived by the user community to derive from a bottomless barrel, they will behave accordingly and will give precious little consideration to the very real need to conserve bandwidth.

Next, a metro network, because it straddles the line between the end user and the network over which the user's traffic will be transported, must be able to offer differentiable levels of QoS to accommodate the many different cost, performance, and service requirements that customers will clamor for.

A metro network must be scalable for the simple reason that the environment it serves will grow and evolve. Metropolitan areas change constantly as they expand, contract, and absorb suburban areas. As a result, a metro design must take into account the fact that the network *will* change and must therefore be built with that eventuality in mind. Furthermore, for economic reasons, the network should be designed to reuse the existing infrastructure to the degree possible, as long as it is not at the expense of service quality. Given the Byzantine depreciation schedules that service providers are often required to abide by for their capital acquisitions, any design that enables in-place capital assets to be fully depreciated will have a sincere advantage over those that fail to take this factor into consideration.

Next, and perhaps most important, the network must be able to survive physically disruptive events. If a router fails, a switch dies, or a fiber route becomes unusable due to a cable cut, the metro network must be able to simply shrug its shoulders, consider a list of options, and continue with business as usual.

Finally, the metro network must be protocol, service, and application independent. Agnosticism is an important construct in modern networks because it speaks to the ongoing convergence phenomenon. Service providers have a strong desire to simplify and therefore reduce the cost and complexity of the network infrastructure they must operate, administer, maintain, and provision. Any design that makes it possible to transport multiple complex traffic types across a single, converged network infrastructure without impact on the end-to-end service quality will be richly rewarded. Consider, for example, the plethora of interfaces and protocols that service providers must contend with. A modern network must be able to interoperate with the following interfaces (and this is by no means an exhaustive list):

- 300 bps *Teletype* (TTY) (yes, it's still around)
- 300 to 1200 bps telemetry
- 9.6 Kbps lottery/ATM machine
- 14.4 to 56 Kbps asymmetric dialup data access
- 56 Kbps symmetric dialup data access
- 64 Kbps TDM voice
- 128 Kbps bonded ISDN
- 128 to 512 Kbps video and DSL
- 1.544 to 2.048 Mbps T1/E1
- 45 Mbps to 40 Gbps optical interfaces

That same network must be able to handle a wide variety of protocols, including *Phase Shift Keying* (PSK), *Differential Phase Shift Keying* (DPSK), *Frequency Shift Keying* (FSK), *Amplitude Shift Keying* (ASK), *Quadrature Phase Shift Keying* (QPSK), *Quadrature Amplitude Modulation* (QAM), *Discrete Multi-tone* (DMT), *Carrierless Amplitude/Phase modulation* (CAP), I.430, *Synchronous Data Link Control* (SDLC), *Link Access Procedure for the D-Channel* (LAPD), Q.931, IP, *Link Access Procedure to Frame Mode Bearer Services* (LAPF), Q.921, ATM, SONET, SDH, *Frequency Division Multiple Access* (FDMA), *Time Division Multiple Access* (TDMA), *Global System for Mobile Communications* (GSM), CDMA, 802.3, 802.5, 802.11, and a host of others. Keep in mind that the network becomes far more diverse and heterogeneous as it gets closer to the customer, and since the metro network often serves as the aggregation point for user traffic before it is classified and handed off to the core, it must be able to handle a wide variety of traffic types and physical interfaces.

Because of the infrastructure that is becoming most common in the metro environment, these networks are often referred to today as *Metro Optical Networks* (MONs). Optical technology has long been considered a transport option for subsea cables and transcontinental networks because of the enormous bandwidth required for coast-to-coast call volumes. Today, because of the burgeoning bandwidth requirements of multimedia-enhanced applications, that same capacity is in demand by metro customers. In Part 2, we begin with a discussion of the MON world.

2

The Metro Network

In this part, we explore the evolution and development of metropolitan networks with a focus on *metropolitan optical networks* (MONs). As we will see, "metro" means different things to different constituencies, so we will make every effort to clearly define the network region where we find metro deployments.

What Is a Metro Network?

As the Internet continues to grow in chaotic and unexpected ways, and as the Internet user base becomes more technologically informed and focused on applications, those very applications undergo innovative evolution as they place growing demands on the network for very high speed, scalable, flexible, and protocol-transparent bandwidth services. At the same time, advances in *Dense Wavelength Division Multiplexing* (DWDM) technology are allowing new network models to be created. Once found only in the long-haul network, DWDM is now evolving so that it provides tremendous bandwidth granularity in the metropolitan and access networks. The result of DWDM's incorporation in the metro has caused the industry to refocus on applications, with a goal to offer the same benefits in that environment as those that have long been available in the long-haul space.

The current literature defines MONs as those that provide connectivity and transport to large metro areas that have a high concentration of business customers and that span distances of no more than 10 miles. As we stated before, they provide the bridge between the long-haul environment and the access environment, and interconnect a wide array of protocols from both enterprise and private customers on access networks to diverse backbones. This seeming disparity of functionality results in a remarkable collection of design, engineering, and deployment challenges due to several factors: the significant installed base of *Synchronous Optical Network/Synchronous Digital Hierarchy* (SONET/SDH) that is common in existing metro area networks, the functional and performance disparities between the access and transport interfaces, and the need to incorporate new technologies and services while protecting legacy offerings, most notably those based on traditional *Time Division Multiplexing* (TDM) technologies such as T1/E1 and SONET/SDH.

These networks were created to transport a relatively limited collection of traffic, specifically voice and private-line services. Those services, largely point-to-point in nature, dominated the technology base for a very long time, but as the metro marketplace began to emerge as a viable standalone entity, they began to show their limitations. The need arose to simplify the network and make it more flexible and quickly provisional because of the nature of the metro customer. At the same time, these networks had to provide volumes of bandwidth, and that bandwidth had to be available more or less instantaneously. T1/E1 certainly offered bandwidth, although SONET and SDH offered vastly more. The problem with SONET and SDH was that although they offered bucketloads of bandwidth, they were not flexible nor were they efficient. The solution lay in the so-called multiservice optical network, which gave service providers the ability to accomplish all the requirements just mentioned: quick provisioning, accurate billing, flexible deployment, low-cost solutions, and greater network intelligence.

All three service regions of the metro marketplace—the service provider, the customer, and the hardware manufacturer—benefit from the migration to the all-optical metro network. The opportunities to provision new customer groups and to offer diverse classes and *quality of service* (QoS) are rich and to a large extent untapped today.

Some players are succeeding in the metro market because they have figured out the relationship between technology and the services made possible by the technology and have directed their efforts accordingly. Those that understand that customer-oriented services drive revenues far more than readily available bandwidth are well positioned, whereas those that have made technology plays without well-thought-out application and solution plans are in peril. Some have already fallen; others will undoubtedly follow. Others are enjoying growing revenue streams and will continue to do so. Companies that create well-targeted, timely, and cost-effective solutions delivered over mesh optical metro networks that support a variety of revenue-generating services will enjoy longevity from the long-term interests of broad service provider markets.

Metro Touch Points

Metro networks sit squarely between the long-haul and access regions of the overall network, as shown in Figure 2-1. We will examine each of these segments because in some ways the lines of functionality between them blur, making it difficult to determine where one ends and the next begins.

Figure 2-1
The networking
hierarchy. Metro
serves as an
intermediate region
between core and
access.

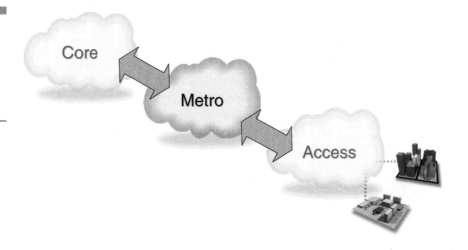

Figure 2-1
The networking hierarchy. Metro serves as an intermediate region between core and access.

Long-Haul Networks

Service providers with national or global scope typically provision long-haul networks. Their networks range from regional (tens of route miles) to global (thousands of route miles) and are used to interconnect disparate metro networks as a way of providing multinational connectivity among a group of regional networking domains. Historically, their principal differentiator has been transport capacity (megabits of provisionable bandwidth) as measured by the capabilities (and limitations) of the underlying transport technology—typically SONET or SDH. To deal with the problems of fiber (and therefore bandwidth) exhaust, most large carriers have deployed DWDM as an effective bandwidth multiplication technique.

The high labor cost involved in the provisioning of long-haul fiber cables resulted in a significant degree of interest in DWDM when it was first announced. DWDM enabled service providers to augment available bandwidth by adding channels rather than by adding fiber, resulting in enormous cost savings. The problem is that as the channel count on a single fiber strand increases, a variety of transmission impairments—crosstalk, if you will—arise and get worse as the channel count climbs. Optical engineers have created clever techniques for dealing with these impairments, including chemically enhanced fibers, optical amplification techniques, and more powerful lasers. Eventually, however, if the distance is too great or too much amplification takes place (resulting in amplified noise along with the desired signal), the optical signal must be converted to electrical, cleaned up, regenerated, converted *back* to optical, and retransmitted.

This complexity translates into a significant expense, the result of which is that service providers contemplating the installation of a long-haul optical network treat it as a strategic, long-term investment. Because of technology innovation, however, long-haul networks are growing in capacity and migrating toward terabit core architectures based on high channel count DWDM. Soon they will offer an additional layer of intelligence that will incorporate wavelength switching and routing, differentiable QoS, and survivability guarantees.

Access Networks

As the name implies, access networks are found at the edge of the MON and typically support a wide array of protocols. Access network bandwidth ranges from subrate services such as DS1, DS3, OC-3/STM-1, OC-12/STM-3, 10 Mbps Ethernet, and 200 Mbps *Enterprise System Connectivity* (ESCON) to higher-speed services such as OC-48/STM-16 and OC-192/STM-48. The customers for these services are as far ranging as the services themselves: residential Internet users, *small office/home office* (SOHO) subscribers, telecommuters, small businesses, large corporations, government agencies, and educational institutions. To support the requirements of such a diverse customer base, these access networks must support a significant number of applications, which in turn require a significant number of access protocols. These include the *Internet Protocol* (IP), *Asynchronous Transfer Mode* (ATM), SONET/SDH, Ethernet, Fast Ethernet, Gigabit Ethernet, 10-Gigabit Ethernet, traditional TDM voice, digital video, *Fiber Distributed Data Interface* (FDDI), *Enterprise System Connection* (ESCON), *Fibre Channel Connectivity* (FICON), and Fibre Channel.

The optical metro market segment is a dynamic, chaotic, and potentially lucrative opportunity for the service provider willing to make a play there. It is driven by evolving end user applications based on high-speed access to databases and high-speed access technologies such as *Digital Subscriber Line* (DSL), cable, fixed wireless, and, eventually, *third generation* (3G). Although the number of transported protocols is large and growing, one fact is key: IP as the protocol of choice is becoming more and more obvious. Because IP is bursty and asymmetric, and it sports a naturally unpredictable behavior, its use poses special challenges to service providers accustomed to the naturally well-behaved nature of the *Public Switched*

Telephone Network (PSTN). Furthermore, new applications have emerged that place large and unpredictable loads on existing networks, including *local area network* (LAN)-based videoconferencing, telemedicine applications, Webcasting, and *Integrated Services Digital Network* (ISDN)-based videoconferencing. Others will emerge over time, and the one characteristic they will share is a growing hunger for bandwidth.

Two key forces drive access networks: evolving applications with diverse service-quality demands and a range of network architectures. Equally important is transparency, otherwise known today by the much overused term *protocol-agnostic*. From a service provider's point of view, transparency is perhaps the most important characteristic of metro networks because it guarantees the ability to build multiprotocol networks.

A phenomenon exists that is well known among electrical engineers called *skin effect*. Skin effect speaks to the fact that current does not flow in the core of a conductor. Instead, it flows on the surface, because of the reduced electrical resistance due to the far greater surface area there. This is why stranded wire is a greater conductor than solid: the multiple strands of wire provide enormously greater surface area than a single solid conductor.

The analogy works well in our discussion of metro networks: Far more "market opportunity surface area" is available at the edge of the network cloud than there is at the core, and although the core marketplace continues to be important, the opportunities, the growth, and the money lie at the edge.

As we discussed in Part 1, "A Lesson in Medieval History," the industry has segmented itself into functional layers that satisfy the changing demands of a diverse application set. A food chain is at work through which the various players combine their efforts in a form of value-added service to satisfy customer demands. The collective efforts of the various manufacturing layers and those of the service providers result in the foundation of what is required to create value for the customer. That value derives from reliable hardware, capable software, targeted applications, and a focus on the customer's value chain.

The Metro Area Itself

Metro networks move traffic within the metro domain, to and from business locations, interoffice facilities, and long-haul network provider points of presence. They are driven both by changing customer requirements and the technology required to support them. They also offer a variety of networking protocols and bandwidth to match those offered by the access region of the network.

Because it is what they know best and have most widely installed, service providers use SONET or SDH for metro transport, relying on either point-to-point connectivity or *add-drop multiplexer* (ADM) ring architectures. These connections are either permanently or semi-permanently connected and range in bandwidth from OC-3/STM-1 to OC-48/STM-16. Because of the complex structure of SONET/SDH, equally complex mappings are required to encode protocols into SONET/SDH frames. Keep in mind the conditions under which SONET and SDH were created: It was a time of telephone monopoly, when traffic largely comprised voice with a bit of data mixed in for color. Networks, traffic, protocols, and applications were far simpler back then than they are now, and the rigid framing structure of legacy SONET and SDH was perfectly fine for the limited requirements of the traffic of the time. As networks and the traffic they move have evolved, however, it has become clear that SONET or SDH do not offer the efficiencies necessary for these evolving networks. Furthermore, because of their rigid framing structures, SONET and SDH do not offer flexible service definitions and suffer from long provisioning cycles.

The Evolving Metro Network

Applications, bandwidth, protocols, QoS, flexibility, service diversity—these drivers are growing and changing in response to the evolving metropolitan environment, which in turn is shifting in response to significant changes in the access and long-haul regions of the network. For example, the remarkable growth in DWDM channel capacity, combined with the dramatic rise in the use of IP in the access domain, has brought sharp focus on metropolitan networks and has made the region fiercely competitive. And although demands have gone up and competition has grown rapidly, the existing legacy TDM infrastructure offers a variety of challenges for service providers looking to meet the efficiency, scalability, and flexibility requirements of customers.

Customer demands have also grown more compelling over time. As applications have evolved, enterprise customers have demanded more and more bandwidth, ultimately rising to demands for OC-48 (STM-16) connections that terminated at the business location on the back side of large IP routers and ATM switches. This was unheard of; line speeds at this level of bandwidth had never been demanded before. Furthermore, it became somewhat clear to service providers and manufacturers alike that the distinction between the access and metro domains of the network were beginning to blur.

To address this ongoing functional evolution, a migration to a scalable, high-capacity architecture was needed for the network. As it dawned on service providers and manufacturers that a key differentiator among competitors was QoS, the functional components of QoS became crucial: bandwidth on demand, service-level differentiation, cost, transparency, and rapid provisioning.

It was clear from the beginning that SONET and SDH could not offer these characteristics at the levels required to maintain a competitive position in the *metropolitan area network* (MAN). As demand grew and tension among the players increased, it became clear that metro-based DWDM was a reasonable solution for the provisioning of high bandwidth. By adding multichannel transport to MONs, service providers found that they could offer flexible, manageable, scalable, protocol-independent services that were also cost effective. This represented the creation of true *multiservice optical networks* (MONs). Most metro players have already made the leap to DWDM, and those that remain will be forced to do so eventually, because DWDM is the most cost-effective answer to high-speed, multichannel provisioning.

Of course, DWDM presents a challenge to service providers. It was primarily designed for long-haul, high-bandwidth, circuit-switched networks where bandwidth was at a premium and where the capability to conserve it or manage it cost effectively was of paramount importance. As demand for metro connectivity has grown, an interesting functional disparity between the legacy circuit-switched networks and the modern packet-based metro networks has emerged. Legacy long-haul networks largely rely on SONET or SDH for high-bandwidth transport; unfortunately, the large ADMs used in that environment do not scale well in the metro arena. Adding DWDM to a SONET/SDH ring may add bandwidth, but it also adds an inordinately high degree of complexity to a network that by all rights should be simple. Although many believe that metro networks are nothing more than long-haul networks built to a Lilliputian scale, they are not. Metro optical networks offer significant capabilities that are designed around the unique requirements of the metro environment. Without these qualities, MONs are not cost effective.

These characteristics include

■ **Diverse protocol support**   Because metro networks are close to the customer, they must be able to support a wide range of both access and transport protocols. Equally important, they must be extremely dynamic and adaptable so that they can keep up with the often unpredictable requirements of the user community. Convergence plays an important role here; the degree to which the network can support diverse payload types from a single deployment platform is a direct measure of the degree to which the network will deliver a competitive advantage to its user community. And because MONs tend to be protocol and device independent, they support legacy installations as well as newly introduced connectivity solutions. This capability to blend the old with the new leads to significant cost reductions due to the elimination of redundant interface equipment and maintenance and provisioning organizations. It also leads to consolidated (and therefore much less complex) network management (by eliminating layers and equipment), simplified network management, and reduced collocation issues.

- **Transparency** Although it sounds like a bad joke, signal transparency at the physical (optical) layer is critically important to multiservice access, metro, or long-haul networks. Because the network tends to be the least common denominator in transport, it must be able to handle a diverse range of protocol types, particularly on the access side. By extending the involvement of optical to the edge of the network, service providers can simplify their networks, reduce provisioning time, limit channel latency issues, and reduce costs because of the growing simplicity of the network. For example, expensive optical-to-electrical transceivers result in significant improvements in scalability as well as drastic reductions in cost. Because optical transport can eliminate the dependency on subrate metallic channel grooming, both cost and operational efficiencies result. It is no wonder that metro providers like Telseon and Yipes have made transparency a key differentiator in their networks. It is also interesting to note that some companies building products for the metro market have recognized the expensive protocol disparity that exists and have done something about it. Consider, for example, the low-cost media converters built by Metrobility (www.metrobility.com).

- **Mesh network support** In a prior book on optical networking, I applauded the merits of companies like Astral Point (www.astralpoint.com), recently acquired by Alcatel, for their innovative bandwidth provisioning and management systems. Figure 2-2 illustrates how they work. A network manager brings up a spider drawing of the network on a management console and clicks the

Figure 2-2
Automated network
provisioning. The
network manager
clicks an origination
point (A) and then
selects the circuit
endpoint (Z). The
circuit is instantly
provisioned at the
appropriate
bandwidth level.

circuit origination point (1). A pull-down menu appears from which the circuit bandwidth is selected (2). A window appears, instructing the network manager to click the circuit endpoint (3). At this point, the bandwidth is provisioned and the circuit is up and running—no waiting and no provisioning interval.

Mesh networks lend themselves to this kind of model because they are designed to provide virtually limitless alternate paths between circuit endpoints. Traditional metro networks have always been based on ring architectures. Recently, however, mesh architecture have begun to appear, challenging the supremacy of rings.

Rings are innovative solutions to network challenges, but they do have limitations. First, in spite of all the capabilities they bring to the network, they are not particularly flexible when it comes to physical deployment. Users are typically required to transport their traffic along the single path that the ring provides, even if that path offers a less than efficient path from the source to the destination. In many cases, this poses no problem, but for the many delay-sensitive applications that are emerging on the scene such as *voice over IP* (VoIP), interactive and distributive video, and others, the added delay can change the perceived QoS, especially if a ring switch has occurred that adds even more end-to-end delay.

In response to this evolving perception, a new architecture has emerged on the scene. Interestingly enough, it is not a new model, but rather one that has been around for a long time. *Mesh architectures* are suddenly enjoying great attention as the next generation of networking to inherit the crown from the pure ring. This model, shown in Figure 2-3, offers the multiple route capabilities of a dedicated network span alongside the survivable nature of a ring. As the figure shows, the bulk of the network is a collection of point-to-point facilities that provide the shortest distance transport between any two end points. However, the mesh can easily be seen to comprise a collection of interconnected rings, and the large number of alternate routes means that survivability is assured alongside guarantees of least-hop routing. In the real world, fiber has now been deployed so extensively and universally that the ability to build mesh networks is no longer a problem.

The real issue with SONET and SDH as they evolve to accommodate the needs of the so-called next-generation network is that both must be modified to transport the variety of data types that have emerged as viable revenue components in the emerging metro optical network. SONET and SDH were designed to address the relatively predictable 64 Kbps transport needs of voice networks that dominated the attention of service providers in the pre-Internet world. However, with the arrival of the World Wide Web in the 1990s, transported traffic became a mix of predictable voice and unpredictable, chaotic, latency-friendly packet data. As a result, SONET and SDH

Figure 2-3
A mesh-ring hybrid network. This architecture offers the resiliency of a ring with the flexibility of a mesh.

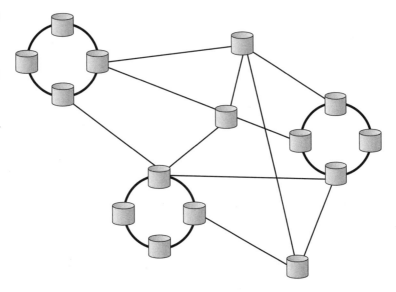

were no longer as capable as they once were for meeting the demands of all service types, particularly given the interest in predictable QoS as the principal differentiator among access and transport providers. Such qualities as security, latency, granular bandwidth provisioning, dynamic time-of-day provisioning, multiple levels of service protection, and a host of others have garnered the attention of service providers in general, particularly as they have begun to managerially segment their networks into local, metro, regional, and long-haul quadrants. It is this differentiable QoS capability that not only provides differentiation among the players in a commodity market, but it also offers new opportunities for revenue generation.

Because of the original services that they were designed to transport, SONET and SDH comprise large, multinode rings that interconnect to other rings as well as to point-to-point transport facilities. These architectures are well understood, fully functional, and widely deployed; thus, the cost of maintaining them on an ongoing basis is comparatively low. They provide ideal carriage for the limited requirements of circuit-switched voice, offering not only low-latency transport but survivability as well. If a failure occurs due to a fiber disruption, service can usually be restored in less than 50 milliseconds, which means that voice traffic is not affected.

As the traffic mix has evolved, however, the limitations of SONET and SDH have become more obvious. Each ring in a SONET or SDH network is limited to a certain number of nodes, and each ring cannot exceed a certain maximum circumference if transport QoS is to be assured. Thus, if added capacity in the network is needed, the solution is to add rings, clearly an expensive and time-consuming process. This stacked ring model also has the disadvantage of being disparately managed, meaning that each ring must be provisioned and managed on an individual basis, making the job of the network manager more difficult.

Furthermore, SONET and SDH do not offer particularly flexible bandwidth allocation capabilities. Originally created to transport 64 Kbps voice, neither of the two adapts well to the transport of data traffic that has wildly variable bandwidth requirements, in terms of both bandwidth assignment and predictability. Think about it: In a SONET network, bandwidth jumps from 51.84 Mbps (OC-1) directly to 155.52 Mbps (OC-3) with no intermediate level. Customers wanting to make a minor upgrade must increase their purchased bandwidth threefold, much of which they will probably not use. Because SONET and SDH rings typically reserve as much as half of their total available bandwidth for redundancy purposes, they are *terribly* inefficient.

The mesh model discussed earlier represents the solution to this set of problems. In the last three years, optical networking has evolved in three significant areas: the development of true all-optical switching and intelligent routing, the extensive proliferation of fiber throughout most carriers' operating areas, and the return of the mesh network.

In a ring network, nodes are connected to one another in such a way that they do not have a direct interconnection to one another; all node-to-node traffic (other than between adjacent nodes) must flow along a rigidly deterministic path from a source node to a destination. In a mesh network, every node in the network is connected to every other node in the network, thus allowing shortest hop routing throughout the network between any two end points.

The advantages of this design are rather strong. First of all, distance limitations in mesh networks are largely eliminated because paths are created on a shortest-path basis between nodes rather than all the way around a ring. As a result, nodes represent the bottleneck in mesh networks rather than the fiber spans themselves, which means that network operators can increase capacity simply by adding nodes on a demand basis and increasing transported bandwidth across the installed fiber infrastructure. This eliminates the stacked ring problem and dramatically improves upgrade intervals, an area of some concern for most service providers.

Perhaps the greatest advantage of mesh networks is management. Unlike SONET and SDH where rings must be managed on a ring-by-ring basis, mesh networks are designed to accommodate point-and-click provisioning, described earlier, which shortens installation intervals from months in many cases to hours. The service focus for these topologies is primarily on the metropolitan marketplace.

The mesh network also enables carriers to design "distributed protection" schemes, which in turn enable them to use their available bandwidth much more efficiently. SONET and SDH rely on redundant rings to provide 100 percent survivability, which on the one hand is great because it provides 100 percent redundancy, but on the other hand wastes 50 percent of the available bandwidth because of the reservation of an entire ring for the eventuality of a rogue backhoe driver with a map of network routes in his back pocket. Because of the tremendous flexibility of mesh topology, entire routes do not need to be reserved. Instead, network designers can allocate backup capacity across a collection of routes, thus preserving the integrity of the network while at the same time using the available bandwidth far more efficiently than a ring.

Furthermore, most carriers have plans to offer a variety of protection levels to their customers as one of many differentiable services. SONET and SDH, with their 100 percent, always-on protection schemes, do not lend themselves to this capability as a mesh architecture, whereas a mesh can be provisioned in any of a variety of ways. These include 0×1 protection (no protection at all), 1:N protection (one protect path that handles the potential failures of multiple paths), and 1:1 protection (dedicated protection à la SONET and SDH rings). 0×1 protection is obviously the most economical, while 1:1 protection is the most costly since it offers 100 percent protection against network failures. Obviously, mesh networks are far more complex than their ring counterparts and are therefore significantly more difficult to manage. They require a level of network management intelligence that is quite a bit more capable than that provided by the SONET and SDH overhead bytes.

Both ring and mesh networks have their advantages and disadvantages, as we have just discussed. Furthermore, SONET and SDH enjoy an enormous embedded base of installed networks and service providers that have made large capital investments to build them and are loathe to discard those investments easily. Not only must hardware be modified, but the evolution from ring to mesh involves modifications to provisioning systems, billing systems, operations support, provisioning, and personnel training as well. The result of this is that most industry analysts believe that the end result is the evolution of hybrid networks that sport the best of both ring and mesh architectures.

Because of the increasingly bursty (and therefore unpredictable) nature of metro area traffic, the ability of network designers and traffic engineers to optimize the network for the loads is proving to be an interesting and often vexing problem. Even though metro networks tend to be relatively small geographically, the amount of traffic generated within them is growing at a rapid clip, and the boundaries of the typical metro network are expanding apace. As a result, service providers are seeking solutions that will augment their existing services and that will extend the service area of legacy infrastructures by taking into account such variables as customer service demographics, variable traffic demands, and time-of-day variations in load. Unfortunately, this type of real-time adaptation of the network requires flexible protocols that can accommodate the requirements of such a technologically chaotic environment.

In most metro networks today, the predominant topology is ring-based and usually depends on SONET/SDH physical layer protocols to frame the transported payload. To add rudimentary flexibility in response to evolving service demands, network designers have in some cases augmented existing ring topologies with DWDM as a bandwidth mediation mechanism. In some network designs, optical ADMs have the capability to selectively "route" individual wavelengths (lambdas) from the source port to the destination port, thus adding yet another layer of flexibility. The problem with these innovations is that they still depend on legacy ring architectures and are therefore an intermediate stage between legacy installations and future network architectures. Generally speaking, *Optical Add-Drop Multiplexer* (OADM)-based network designs require significant, complex network pre-planning and are not particularly flexible. Instead, they offer massive amounts of bandwidth, which to a certain extent can substitute for flexibility.

Although these hybrid arrangements have become somewhat popular as a transition technology, in reality they are not particularly scalable. SONET/SDH rings are rather rigid in terms of the way in which they manage on-ring bandwidth and therefore do not scale well when placed into service with other rings.

One response to this collection of challenges is DWDM-enhanced mesh networks. Both carriers and hardware providers have developed interest in this combination because it takes best advantage of a network model that is effectively a wavelength-switching design, enhanced by the flexibility of mesh connectivity. This leads to the next generation of network device intelligence in the form of the *optical cross-connect* (OXC). In OXC installations, network paths can be established rapidly and are both protocol and bandwidth independent. Furthermore, they can be used in both point-to-point configurations and in ring-based ADMs. Because the dominant protocols that will ride over these optical connections are (and will continue to be) ATM and IP, both of which rely on mesh networks to achieve their significant advantages, mesh networks represent a more sensible approach to interface with these protocols at layers two and three.

Mesh networks are more significantly more bandwidth efficient than ring-based architectures. To service providers, they represent a topology that they are quite familiar with and that requires little in the way of up-front provisioning, since all "routes" are preestablished by the very nature of the physical network. Furthermore, they scale well, giving the network provider some semblance of control over their incremental costs.

In short, mesh networks offer quite a few advantages, most of them related to their capability to create a smooth transition from legacy infrastructures toward a future network design. As an example, legacy rings equipped with traditional OXCs can be used to offer standard SONET/SDH add-drop functions, but they can also be reconfigured when appropriate to create a "mesh of rings" during the migration stage. Mesh networks provide the following advantages:

- **Rapid provisioning** In legacy SONET/SDH networks, provisioning is a complex, multistage process that commonly takes weeks and is often fraught with errors. Furthermore, the less flexible nature of

SONET/SDH does not allow for "quick and easy" provisioning activities. Because the metro environment has become so competitive, tools have been developed that are software-based and that facilitate rapid, automated provisioning.

In mesh networks that are enhanced with DWDM, the individual wavelengths and the connection points (switching, add-drop, wavelength management) are the resources that must be managed. To ensure that the network is as efficient as possible and to protect (and hopefully augment) service provider revenues, intelligent management processes for wavelength provisioning are often deployed. These capabilities result in very fast connection intervals and manage such functions as end-point identification, route computation, and accurate, low-latency route setup. They also manage such parameters as delay, traffic priority, link protection, and channel quality as a way to deliver QoS.

As IP traffic continues to expand in volume relative to circuit-switched traffic, the cross-sectional profile of network resource utilization will change, and will become significantly more unpredictable and bursty than that of its predecessor. As a result, it makes sense to use existing IP-based route control protocols within these optical networks. *Multiprotocol Label Switching* (MPLS) is the first of these to be used as a "shim" protocol between IP and ATM in the protocol stack, with the intent to eventually displace ATM where feasible; in the optical environment, *multiprotocol lambda switching* (MPλS) has been designed to accommodate the unique requirements seen in the MON environment. A network that is "data-aware" (to use a common expression) must use IP traffic-engineering to manage such disparate tasks as routing and protection. It must also provide manageable integration of IP with the DWDM sublayer for effective bandwidth utilization. One key advantage is that the so-called short-reach optics deployed on modern high-speed routers can be connected directly to DWDM cross-connect devices such as the Lucent LambdaRouter, allowing more effective automated traffic engineering. These *user-to-network interfaces* (UNI) enable metro providers to design highly responsive IP networks, including the one performance indicator that is most visible to the customer, provisioning intervals. They translate into greater market share, increased revenue, improved customer satisfaction, and lower overall costs.

■ **Network survivability** Because of the broad, diverse range of protocols, applications, and network-based services used by the various communities of users within the metro space, and because of the granular (and varied) demand for differing levels of service quality among various customer groups, service providers must craft different QoS and network performance criteria for those different customers. *Service level agreements* (SLAs), particularly for voice and video services, have once again become a major area for discussion. Even though the bulk of all packets transported across the Internet consists of short, *Transmission Control Protocol* (TCP) *acknowledgement* (ACK) and *nonacknowledgement* (NAK) packets, a variety of delay-insensitive customer traffic components still make up a reasonable percentage of the overall traffic model, including IP, fax, e-mail, web pages, and so on. As a result, some customers will be perfectly happy to put up with slightly lower quality service in exchange for a smaller invoice at the end of the billing cycle.

As we observed earlier, both ring and mesh topologies will continue to coexist in the metro, although mesh networks will dominate over time. Topologically, mesh networks are really nothing more than a collection of nested rings (see Figure 2-3). As with ring networks, survivability comes in two forms: restoration and protection. Survivability functions are classified into the same two forms as well. Restoration refers to real-time, dynamic functions that take place after a network fault and that serve to recover from the fault. They tend to be reactive mechanisms. Protection, on the other hand, refers to those proactive measures taken to prevent failures from mesh networks that can match SONET/SDH 50-millisecond recovery intervals and have the capability to recover from multiple fiber cuts or node faults. Because of the varying demands from customers for bandwidth and QoS, mesh topologies are ideally positioned. They can provide dedicated paths for delay-sensitive traffic or shared resources for those traffic components that are less sensitive to delay.

■ **Network management and OSS** *Operations Support Systems* (OSSs) provide three key areas of functionality: billing and customer care, assurance, and fulfillment. Billing and customer care include such factors as fraud prediction and management, billing mediation, bill rating, tariff management, and interactive customer care. Assurance handles security, trouble tracking and management, performance tracking, service management, and fault identification, tracking, and management. Finally, fulfillment manages flow-through provisioning, inventory, service activation, access and transport engineering, OSS interoperability, network policy management, and voice provisioning.

Most service providers today have historically focused their efforts on increasing the capacity of their networks rather than on improving operational efficiencies and tend to view network management and OSS functions as a necessary evil expense that is one of the most onerous costs of doing business. What the service providers *don't* realize is that well over half of their total cost of doing business can be positively affected through reasonable and a well-thought-out investment in network management tools and OSSs. This is as true in the evolving metro world as it is in voice and legacy data. Today, a variety of functional areas are poised to grow in the OSS domain. These include customer self-management, network resource management, converged service billing, and on-demand services and provisioning. In 2001, service providers spent $9 billion on OSS services and software, and the numbers are expected to grow.

These systems offer a number of distinct advantages to the service provider. More than anything else, they enable the service provider to operate much more efficiently, which in turn translates into reduced costs and higher revenues; these translate into improved profits.

In spite of the new protocols and services that the market is demanding, service providers require the same capabilities that they always have: standards-based management systems, bit-rate and protocol-independent management systems, support for integrated technology solutions, and the capability to monitor performance, fault isolation, and accounting activities, all of which fall under the broad umbrella of *Operations, Administration, Maintenance, and Provisioning* (OAM&P).

■ **Subrate (shared lambda) provisioning** Although the demand for high-speed optical transport continues to grow, metro network service providers must also provision lower-bandwidth solutions for those customers that do not require high-bandwidth solutions. These include such services as DS-3, OC-3, OC-12, 200 Mbps ESCON, and a variety of others. This creates a challenge for the provider because although optical transport technologies offer a high degree of bandwidth granularity, legacy technologies such as SONET and SDH really do not. If an entire wavelength must be assigned to a customer to satisfy low-bandwidth demands, the approach is clearly wasteful. It is significantly more efficient to provision fractional wavelengths and to aggregate them as required into a single channel. Both data traffic and legacy circuit traffic can be transported in the same lambda at the ingress side, thus dramatically improving the efficiency of the overall system.

This also has another positive impact on the service provider; it enables them to "pay as you grow" so that they can intelligently augment the network as it makes financial sense, rather than doing an entire buildout from day one.

- **Reliability**   Service interruptions in MONs can cause serious service disruptions because of the numbers of users that exist on the networks. The high concentration of disparate customers found in MTUs can result in widespread damage if a network fails in the metro environment. As a result, MONs must provide high reliability if they are to be taken seriously.

- **Cost**   DWDM technology becomes important in the metro space because it has the capability to consolidate traffic—and therefore cost—onto a single shared facility that offers appropriate service granularity to the customer as required. Cost can rapidly become a major issue, because for the most part these companies already have a significant investment in SONET/SDH infrastructure, which has already proven itself with tremendous economies of scale. The real benefit in the metro arena comes when service providers, in concert with hardware vendors, offer end-to-end solutions.

Metro Demographics

For a very long time, the metro geography was dominated by large corporations that occupied single-tenant buildings. Today that model has changed drastically. Most corporations in major metro areas now occupy a large number of smaller buildings that are located at the periphery of the metro area (where the customers are) instead of a single monolithic facility. Of course, the employees in those far-flung facilities require that their connectivity solution be functionally seamless, such that there is no service difference between the network they had when they were all in the same building and the network they now have while scattered across the landscape. This means that metro technologies must be up to the task of delivering this degree of "seamlessness."

Although numerous legacy technologies can offer the level of connectivity required, they tend to be expensive because they are designed for large installations across which the cost of deployment can be amortized. As a result, they do not always scale well economically when deployed in metro networks. Other solutions, discussed in Part 3, provide a better answer to metro connectivity.

Regulatory Impacts

The "reworking" of the telecommunications industry that began with the divestiture of AT&T in 1984 reaches far beyond the borders of the United States. It has become a global phenomenon as companies move to compete in what has become an enormously lucrative business. Customers, however, have been quick to point out that technology is a wonderful thing, but unless the provider positions it in a way that makes it useful to the would-be customer, it has very little value. Customers expect service provider representatives to have relevant technical knowledge as well as knowledge about them as customers: business issues, competition, major segment concerns, and so on.

Furthermore, the level of customer satisfaction or dissatisfaction varies from industry segment to industry segment. *Incumbent Local Exchange Carriers* (ILECs) are too rigid and expensive. *Competitive Local Exchange Carriers* (CLECs) must compete on price, but they must also show a demonstrated ability to respond quickly and offer OSS/*Network Management* (NM) capabilities as well as, show staying power.

Ninety-two percent of all business customers want to work with an account team when purchasing services, yet account teams are not creating long-lasting relationships or loyalty in their day-to-day dealings with customers. Eighty-eight percent of those customers buy their services from the local provider, but only 53 percent say that they would continue to use them if given the choice. According to numerous studies, the single most important factor that customers take into consideration when selecting a service provider is *cost*. Yet the most important factor identified for improvement is *customer service.*

The most daunting challenge that has historically faced service providers is the requirement to provide true universal service, both in rural areas as well as in metropolitan and suburban areas. The American Communications Act of 1934, signed into law by President Roosevelt, mandated that telephony service would be universal and affordable. To make this happen, AT&T agreed in the 1940s to offer low-cost service with the tacit approval of the Justice Department to charge a slightly higher price for long distance, some business-line services, and value-added services as a way to offset the cost of deploying high-cost rural service and to support low-income families. These subsidies made it possible to offer true universal, low-cost service.

Today, 60 to 70 percent of the local loops in the United States are still subsidized to the tune of anywhere from $3 to $15 per month. As a result, the incumbent service providers often charge significantly less for the service they sell than it actually costs them to provide it.

The problem with this model is that only the ILECs have the right to enjoy these subsidies, which means that CLECs are heavily penalized right out of the starting gate. Couple that with the fact that 40 percent of all CLEC revenues go to the ILECs for access charges. As a result of these and other factors, CLECs typically ignore the residence market in favor of the far more lucrative business markets. Business markets are significantly easier and less costly to provision than residence installations because of the dominance of *multitenant buildings* (MTUs). Clearly, the metro market is a desirable place to be. Of course, there is an added issue: ILECs control more than 60 percent of all metro spending.

The regulatory environment in many countries fails to address the disparity that exists between incumbent providers and would-be competitors. Recent decisions, such as the United States' Communications Act of 1996, consisted of a series of decisions designed to foster competition at the local loop level. Unfortunately, by most players' estimations, it has had the opposite effect.

New technologies always convey a temporary advantage to the first mover. Innovation involves significant risk; the temporary first mover advantage allows them to extract profits as a reward for taking the initial risk, thus allowing innovators to recover the cost of innovation. As technology advances, the first mover position often goes away, so the advantage is fleeting. The relationship, however, fosters a zeal for ongoing entrepreneurial development.

Under the regulatory rule set that exists in many countries today, incumbent providers are required to open their networks through element unbundling and to sell their resources—including new technologies—at a wholesale price to their competitors. In the minds of regulators, this creates a competitive marketplace, but it is artificial. In fact, the opposite happens; the incumbents lose their incentive to invest in new technology, and innovation progresses at "telco time." Under the wholesale unbundling requirements, the rewards for innovative behavior are socialized, while the risks undertaken by the incumbents are privatized. Why should they invest and take a substantial economic risk when they are required by law to immediately share the rewards with their competitors?

Ultimately, the truth comes down to this: Success in the access marketplace, translated as sustainable profits, relies on network ownership—period. Many would-be competitors such as E.spire and ICG bought switches and other network infrastructure components, built partial networks, and created business plans that consciously relied on the ILECs for the remainder of their network infrastructure. Because of the aforementioned subsidies, they quickly learned that this model could not succeed, and many of them have disappeared.

Consider that when WorldCom bought *Metropolitan Fiber Systems* (MFS), they weren't after the business; they were after the in-place local network that MFS had built, which would enable WorldCom to satisfy customer demands for end-to-end service without going through the cost of a network build-out. They paid more than six times the value of MFS' in-place assets for the company. The message? Network ownership is key.

Two major bills currently before the U.S. Congress (Tauzin-Dingell, HR 1542, and Cannon-Conyers, HR 1697) aim to reposition the players. 1542 makes it easier for ILECs to enter the long-distance, Internet, and data businesses. 1697 tightens controls over their entry and applies significant fines for failure to open incumbent networks to competitors. 1542 has already passed the Senate, but a major battle is expected in the House of Representatives. Most pundits believe that the industry needs less government controls and more competition; the FCC appears to agree.

Many believe that high-speed data and Internet access should have been exempted from the Communications Act of 1996's mandates and that it should have only targeted traditional voice services where the ILECs have clear monopoly positions. They contend that unbundling and wholesale requirements should not be required for nonvoice services.

Recent decisions on the part of the FCC aim to bring about the changes that are needed. Both cable data and DSL are on their way from being classified as telecom services to being reclassified as information services by the agency, a move that would exempt them from 271 unbundling requirements.

Likely Regulatory Solutions

So what are the possible solutions? One is to eliminate the local service subsidies and enable the ILECs to raise rates so that they are slightly above the actual cost of provisioning service. Subsidies could continue for high-cost rural areas and low-income families (roughly 18 percent of the total) but would be eliminated elsewhere. New entrants would then have a greater chance of playing successfully in the local access business. The subsidy dollars, estimated to be in the neighborhood of $15 billion, could then be redeployed to finance universal broadband access deployment. The monies could be distributed among ILECs, CLECs, cable companies, *Internet service providers* (ISPs), wireless companies, and DSL providers to facilitate broadband.

A second solution is to call for the structural separation of the ILECs, which would result in the establishment of retail and wholesale arms. The retail arm would continue to sell to traditional customers, while the wholesale arm would sell unbundled network resources to "all comers." The result would be enhanced innovation; the downside would undoubtedly be strong resistance from the ILECs.

The new regime at the FCC, led by Michael Powell, has indicated that it wants less government intervention in the telecommunications marketplace rather than more, a good thing in light of the current industry. One could argue that well-intended regulatory strictures have in fact done damage.

Consider the case of WorldCom. The company's original plan in the late 1990s was to challenge local service providers all over the world by creating a broadband voice and data IP network through acquisitions and mergers. The regulators, concerned by WorldCom's aggressive plans, felt that the intended company looked too much like a monopoly. They forced the divestiture of MCI's Internet company to Cable & Wireless and rejected the proposed merger with Sprint because of fears that they would control 80 percent of the long-distance market. This decision was made while long-distance revenues were plummeting due to the influence of such disruptive technologies as multichannel optical transport and IP. The result is two badly weakened companies that have not yet recovered—and may not. In fact, they could be prime acquisition targets by the ILECs.

Another example is AT&T itself. AT&T was expected to be a big winner in the local broadband access game following its acquisitions of cable properties for its plans to deliver high-speed Internet and interactive services. Many analysts expected a market cross-invasion between the ILECs and cable providers, but it never happened. Cable providers concentrated on adding Internet service and additional channels; telephone companies concentrated on penetrating the long-distance market. Furthermore, when talk of open access and loop unbundling began to be targeted at the cable industry in 2000, AT&T's hopes of a competitive advantage through cable ownership were dashed.

ILECs were also expected to work hard to penetrate each other's markets, but this never happened either. Who better than the ILECs knows that network ownership is the most critical factor for success in the local access game? If you control the network, you control the customers. More importantly, if you don't have that control, don't get into the game.

In the metro market, which is highly dependent on high-speed access technologies such as DSL (and to a lesser degree, cable and wireless), these regulatory decisions represent pivotal points in the evolution of the metro domain. If the FCC chooses to reduce the regulatory pressure on the incumbents, customers will invest in bandwidth at the edge, driving traffic into the metro, which will in turn drive traffic into the network core.

Summary

Metro networks have evolved over a relatively short period of time in response to changing corporate topologies and demographics, different customer requirements, shifts in network economics, and modifications of regulatory policy. Companies have recognized that the closer they are to their customers, the more effectively they can deliver service to those customers. As a result, specialized metro networks have evolved that satisfy the unique demands in this odd region that lies between the access networks and the network core.

In Part 3, we continue our discussion of metro, this time with a focus on the five technology components that make up the metro arena: customer premises, access, transport, switching, and interfaces.

Enabling Technologies

In this part, we examine the technologies that enable the metro networking environment. We divide these into five categories and will examine each in turn: premises, access, transport, switching, and interfaces.

Premises technologies are those used by the customer for intracorporate communications such as *local area networks* (LANs, most notably Ethernet). In the metro networking arena, premises technologies play an important role because they generate and terminate the content that ultimately drives the demand for bandwidth at the edge of the network. Today the single biggest driver at the premises level is Ethernet. Described briefly in an earlier part of the book, Ethernet has cast off its LAN limitations and is now running freely about the network, performing all sorts of tasks for which it was never intended—yet doing them exceptionally well. It has been so successful as a point-to-point transport scheme that a new type of carrier has emerged, offering carrier-class Ethernet for long-haul and metro applications. We begin our discussion, then, with Ethernet.

Introduction

Ethernet is the world's most widely deployed LAN technology. Originally, the commercial version of Ethernet was designed to support 10 Mbps transport within a LAN. In the last few years, however, 10 Mbps has been supplemented with both 100 Mbps (Fast Ethernet) and Gigabit Ethernet versions, and in recent months a 10 Gbps version has been introduced. Legacy Ethernet was designed to operate over various forms of twisted pair, coaxial cable, and fiber. Furthermore, both bus and star topologies were supported, a characteristic that continues today. As we mentioned earlier, Ethernet often uses a *Medium Access Control* (MAC) protocol called *Carrier Sense Multiple Access with Collision Detection* (CSMA/CD).

Overview and Terminology

In the 20 years since Ethernet first arrived on the scene as a low-cost, high-bandwidth 10 Mbps LAN service for office automation applications, it has carved a niche for itself that transcends its original purpose. Today it is used in LANs, *metropolitan area networks* (MANs), and *wide area networks* (WANs) as both an access and transport technology. Its bandwidth capabilities have evolved to the point that the original 10 Mbps version seems almost embarrassing; a 10 Gbps version of the technology now exists.

Today's Ethernet product is based on the *Institute of Electrical and Electronics Engineers* (IEEE) 802.3 standard and its many variants, which specify the CSMA/CD access scheme (explained in a moment) and a handful of other details. It is perhaps important to note that Ethernet is a product name, originally invented and named by 3Com founder Bob Metcalfe. CSMA/CD is the more correct name for the technology that underlies Ethernet's functionality and the 802.3 standard. The term Ethernet, however, has become the de facto name for the technology.

Ethernet evolved in stages, all based upon the original IEEE 802.3 standard, which passed through various evolutionary phases of capability including 10 (Ethernet), 100 (Fast Ethernet), 1000 (Gigabit Ethernet), and 10,000 Mbps (10 Gbps Ethernet) versions. All are still in use, which allows for the design and implementation of tiered corporate Ethernet networks. A corporation, for example, could build a gigabit backbone, connected to Fast Ethernet access links, and connected in turn to 10 Mbps switches and hubs that communicate with desktop devices at the same speed.

In its original form, Ethernet supported point-to-point connectivity as far as 0.062 miles over *unshielded twisted pair* (UTP), up to 1.2 miles over multimode fiber, and up to 3.1 miles over *single-mode fiber* (SMF) at a range of speeds. However, when the metro market began to emerge as a "newly discovered network region," it began to be seen as an alternative transport scheme for both inter- and intrametro network architectures. Today, 10 Mbps is inadequate for many enterprise applications, so the new versions of Ethernet mentioned earlier have evolved and become standardized.

How Ethernet Works

Ethernet is called a *contention-based LAN*. In contention-based LANs, devices attached to the network vie for access using the technological equivalent of gladiatorial combat. If a station on the LAN wants to transmit, it does so, knowing that the transmitted signal *could* collide with the signal generated by another station that transmitted at the same time, because even though the transmissions are occurring on a LAN, some delay still takes place between the time that both stations transmit and the time they both realize that someone else has transmitted. This collision results in the destruction of both transmitted messages. If a collision occurs, both stations stop transmitting, wait a random amount of time, and try again. This technique is called *truncated binary exponential backoff* and will be explained later.

Ultimately, each station *will* get a turn to transmit, although how long they have to wait is based on how busy the LAN is. Contention-based systems are characterized by *unbounded delay,* because no upward limit exists on how long a station may wait to use the shared medium. As the LAN gets busier and traffic volume increases, the number of stations vying for access to the shared medium, which only enables a single station at a time to use it, also goes up, which results in more collisions. Collisions result in wasted bandwidth so LANs do everything they can to avoid them.

The protocol employed in traditional Ethernet is CSMA/CD. In CSMA/CD, a station observes the following guidelines when attempting to transmit. Remember the old western movies when the tracker would lie down and put his ear against the railroad tracks to determine whether a train is coming? Well, CSMA/CD is a variation on that theme. First, it listens to the shared medium to determine whether it is in use. If the LAN is available (not in use), the station begins to transmit but continues to listen while it is transmitting, knowing that another station could transmit simultaneously. If a collision is detected, both stations back off and try again. To avoid a repeat collision, each station waits a random period of time before attempting a retransmission.

LAN Switching

To reduce contention in CSMA/CD LANs, LAN switching was developed, a topology in which each station is individually connected to a high-speed switch, often with *Asynchronous Transfer Mode* (ATM) on the backplane. The LAN switch implements full-duplex transmission at each port, reduces throughput delay, and offers per-port rate adjustment. The first 10 Mbps Ethernet LAN switches emerged in 1993 followed closely by Fast Ethernet (100 Mbps) versions in 1995 and Gigabit Ethernet (1,000 Mbps) switches in 1997. Today they are considered the standard for LAN architectures, and because they use star wiring topologies, they provide high levels of management and security to network administrators.

Ethernet Services

Ethernet has been around for a long time and is understood, easy to use, and trusted. It is a low-cost solution that offers scalability, rapid and easy provisioning, granular bandwidth management, and simplified network management due to its single protocol nature. As its popularity has increased and it has edged into the WAN environment, it now supports such applications as large-scale backup, streaming media server access, Web hosting, *application service provider* (ASP) connectivity, *storage area networks* (SANs), video, and disaster recovery.

Another important service is the *virtual LAN* (VLAN). Often characterized as a *virtual private network* (VPN) made up of a securely interconnected collection of LANs, a number of vendors now offer VLANs as a service, including Lantern Communications, which claims to be able to support thousands of VLANs on a single metro ring. Similarly, *transparent LAN* (TLAN) services have gained the attention of the industry; they enable multiple metro LANs to be interconnected in a point-to-point fashion.

Many vendors are now talking about *Ethernet private lines* (EPLs), an emerging replacement for legacy dedicated private lines. Implemented over *Synchronous Optical Network/Synchronous Digital Hierarchy* (SONET/SDH), the service combines SONET/SDH's reliability, security, and wide presence with Ethernet's rapid, flexible provisioning capability. One company that sees the value in this combination is Appian Communications, already building *Systems Network Architecture* (SNA), video, SAN, and *Voice over IP* (VoIP) application support across its network architectures.

Of course, Ethernet presents some disadvantages. These include the potential for packet loss and the resultant inferior *quality of service* (QoS). However, two new standards have emerged that will go a long way towards eliminating this problem. The IEEE 802.17 *Resilient Packet Ring* (RPR) standard, discussed later in this book, is designed to give SONET/SDH-like protection and reliability to Ethernet, using a data link layer protocol that is ideal for packet services across not only LANs, but MANs and WANs as well. Primarily targeted at voice, video, and IP traffic, it will restore a breached ring within 50 milliseconds. The IEEE began the 802.17 RPR standards project in December 2000 with the intention of creating a new MAC layer. Fiber rings are widely used in both MANs and WANs; these topologies, unfortunately, are dependent on protocols that are not scalable for the demands of packet-switched networks such as IP. The 802.17 RPR working group promotes RPR as a technology for device connectivity. Issues currently being addressed are bandwidth allocation and throughput, deployment speed, and equipment and operational costs.

The second standard, ITU X.86, maps Ethernet (*Open Systems Interconnection* [OSI] layer two) frames into SONET/SDH physical layer (OSI layer one) frames for high-speed transmission across the broad base of legacy networks. Both standards are due to be ratified in late 2002 or early 2003.

Ethernet: A Brief History

Ethernet was first developed in the early 1970s by Bob Metcalfe and David Boggs at the *Xerox Palo Alto Research Center* (Xerox PARC). Originally designed as an interconnection technology for the PARC's many disparate minicomputers and high-speed printers, the original version operated at a whopping 2.9 Mbps. Interestingly, this line speed was chosen because it was a multiple of the clock speed of the original Alto computer. Created by Xerox Corporation in 1972 as a personal computer targeted at research, Alto's name came from the Xerox Palo Alto Research Center, where it was originally created. The Alto was the result of developmental work performed by Ed McCreight, Chuck Thacker, Butler Lampson, Bob Sproull, and Dave Boggs, engineers and computer scientists attempting to create a computer small enough to fit in an office but powerful enough to support a multifunctional operating system and powerful graphics display.

In 1978, Xerox donated 50 Alto computers to Stanford, MIT, and Carnegie Mellon Universities. These machines were quickly adopted by campus user communities and became the benchmark for the creation of other personal computers. The Alto comprised a beautiful (for the time) graphics display, a keyboard, a mouse that controlled the graphics, and a cabinet that contained the *central processing unit* (CPU) and hard drive. At $32,000, the thing was a steal.

The concept of employing a visual interface (*what you see is what you get* [WYSIWYG]) began in the mid-70s at Xerox PARC, where a graphical interface for the Xerox Star system was introduced in early 1981. The Star was not a commercially successful product, but its creation led to the development of the Apple Lisa in 1983 and the Macintosh in 1984.

In July 1976, Metcalfe and Boggs published "Ethernet: Distributed Packet Switching for Local Computer Networks" in *Communications of the Association for Computing Machinery*. A patent followed with the mouthful of a name, "Multipoint Data Communications System with Collision Detection," issued to Xerox in December 1977.

Phase Two: Xerox and DEC

In 1979, DEC, Intel, and Xerox Corporations standardized on an Ethernet version that any company could use anywhere in the world. In September 1980, they released Version 1.0 of the DIX standard, so named because of the initials of the three founding companies. It defined the so-called thick Ethernet or Thicknet version of Ethernet (10Base5), which offered a 10 Mbps CSMA/CD protocol. It was called Thicknet because it relied on relatively thick coaxial cable for interconnections among the various devices on the shared network. The first DIX-based controller boards became available in 1982, and the second (and last) version of DIX was released in late 1982.

In 1983, the IEEE released the first Ethernet standard. Created by the organization's 802.3[1] Committee, *IEEE 802.3 Carrier Sense Multiple Access with Collision Detection (CSMA/CD) Access Method and Physical Layer Specifications* was basically a reworking of the original DIX standard with changes made in a number of areas, most notably to the frame format. However, because of the plethora of legacy equipment based on the original DIX specification, 802.3 permitted the two standards to interoperate across the same Ethernet LAN.

The ongoing development and tightening of the standard continued over the next few years. In 1985, IEEE released 802.3a, which defined thin Ethernet, sometimes called cheapernet (officially 10Base2). It relied on a thinner, less expensive coaxial cable for the shared medium that made installation significantly less complex. Although both media offered satisfactory network performance, they relied on a bus topology that made change management problematic.

Two additional standards appeared in 1987. 802.3d specified and defined what came to be known as the *Fiber Optic Interrepeater Link* (FOIRL), a distance extension technique that relied on two fiber optic cables to extend the operational distance between repeaters in an Ethernet network to 1,000 meters. IEEE 802.3e defined a 1 Mbps transmission standard over twisted pair, and although the medium was innovative, it never caught on.

1990 proved to be a banner year for Ethernet's evolution. The introduction of 802.3i 10BaseT standard by the IEEE made it possible to transmit Ethernet traffic at 10 Mbps over Category 3 UTP cable. This represented a major advance for the protocol because of UTP usage in office buildings. The result was an upswing in the demand for 10BaseT.

With the expanded use of Ethernet came changes in the wiring schemes over which it was deployed. Soon, network designers were deploying Ethernet using a star wiring scheme (see Figure 3-1) instead of the more common bus architecture, which was significantly more difficult to manage, install, and troubleshoot.

---

[1]A bit of history: The IEEE 802 committee is so named because it was founded in February 1980.

Figure 3-1
A star-wired LAN.
Each station is
individually
connected to the
LAN switch (top) so
that contention is
eliminated, resulting
in a far more efficient
network.

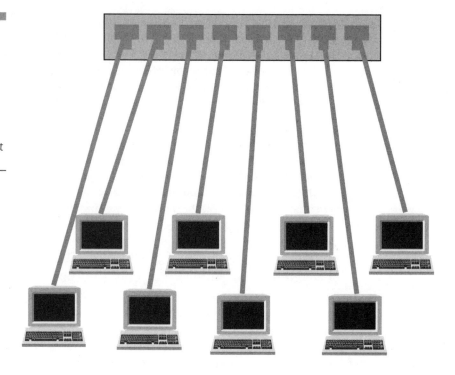

In 1993, the IEEE released 802.3j (10BaseF), a technological leap for Ethernet that extended the transmission distance significantly by stipulating the use of a pair of optical fibers instead of copper facilities. With this technique, transmission distances could be extended to 2,000 meters, a distance that dramatically expanded Ethernet's utility in office parks and in *multitenant units* (MTUs). 802.3j represented an augmentation and update of FOIRL, mentioned earlier.

The Arrival of Fast Ethernet

In 1995, the IEEE announced the 100 Mbps 802.3u 100BaseT standard. Commonly known as *Fast Ethernet*, the standard actually comprises three disparate substandards, which in turn support three different transport media. These include 100BaseTX, which operates over two pairs of CAT 5 twisted pair; 100BaseT4, which operates over *four* pairs of CAT 3 twisted pair; and 100BaseFX, which operates over a pair of multimode optical fibers. It was deemed necessary to offer a version for each of these media types because of the evolving customer environment; more on this will be covered later.

Two-Way Ethernet

"Due to popular demand" as they say, the IEEE announced 802.3x in 1997, which provided a standard that governed full-duplex Ethernet. By allowing stations on an Ethernet LAN to simultaneously transmit and receive data, the need for a medium contention mechanism such as CSMA/CD was eliminated, allowing two stations to transmit simultaneously over a shared point-to-point facility. The standard was written to accommodate all versions of Ethernet, including Fast Ethernet.

The IEEE also released the 802.3y 100BaseT2 standard in 1997, which provided a mechanism for transmitting at 100 Mbps over two pairs of CAT 3 balanced wire.

The Arrival of Gigabit Ethernet

In 1998, *Gigabit Ethernet* arrived on the commercial scene. Like Fast Ethernet, the 802.3z 1000BaseX standard supported three different transmission mechanisms with three substandards: 1000BaseSX relied on an 850 nm laser over multimode fiber, 1000BaseLX used a 1300 nm laser over both single and multimode fiber, and 1000BaseCX was designed for operation over short-haul copper (sometimes called twinax) *shielded twisted pair* (STP). The IEEE also released 802.3ac in 1998, a standard that defines the extensions required to support VLAN tagging on Ethernet networks, mentioned earlier in this section.

The advances continued. In 1999, 802.3ab 1000BaseT hit the market, supporting 1 Gbps transmission over four pairs of CAT 5 UTP.

The Ethernet Frame

Ethernet is a layer two data link standard—in effect, a switching technique. The standard transmission entity at layer two is the frame. The frame is nothing more than a "carriage" mechanism for the user's data, which is usually a packet. The frame, then, has a field for user data as well as a series of fields designed to carry out the responsibilities of the data link layer. These responsibilities include framing, addressing, transparency control, and bit-level error detection and correction. The Ethernet frame, shown in the following minitable, is designed to accomplish its task well. Each field of the frame is described in detail later.

Preamble (7 bytes)	Start Frame Delimiter (1 byte)	Destination MAC Address (6 bytes)	Source MAC Address (6 bytes)	Length/ Type (2 bytes)	MAC Client Data (0-n bytes)	Pad (0-p bytes)	Frame Check Sequence (4 bytes)

The *Preamble* field is a seven-byte sequence of alternating ones and zeroes, designed to signal the beginning of the frame and to provide a degree of synchronization as receiving devices detect the presence of a signal and prepare to process it correctly.

The *Start Frame Delimiter* field comprises a specific sequence of eight bits with the 10101011. This field signifies the functional beginning of the frame of data.

The *Destination MAC Address* field identifies the station that is to receive the transmitted frame. *The Source MAC Address* field, on the other hand, identifies the transmitting station. In 802.3, these address fields can be either two bytes or six bytes in length, although most networks today use the six-byte option. The Destination Address field can specify either an individual address targeted at a single station or a multicast address targeted at a group of stations. If the destination address consists of all "one" bits, then the address is a broadcast address and will be received by all stations on the LAN.

The *Length / Type* field is used to indicate either the number of bytes of data contained in the MAC Client Data field or the type of protocol carried there. If the value of the field is less than or equal to 1,500, then it indicates the number of bytes in the MAC Client Data field. If it is greater than or equal to 1,536, then the field indicates the nature of the transported protocol.

The *MAC Client Data* field contains the user data transmitted from the source to the destination. The maximum size of this field is 1,500 octets. If it is less than 64 bytes, then the Pad field that follows it in the frame is used to bring the frame to its required minimum length.

The *Pad* field enables a "runt frame" to be augmented so that it meets the minimum frame size requirements. When used, extra bytes are added in this field. The minimum Ethernet frame size is 64 bytes, measured from the Destination MAC Address field through the frame check sequence.

The *Frame Check Sequence* field provides a four-byte *cyclical redundancy check* (CRC) calculation for error checking the content of the frame. As with all layer two error control schemes, the originating station creates the MAC frame and then calculates the CRC before transmitting the frame across the network. The calculation is performed on all the fields in the frame from the destination MAC address through the Pad fields, ignoring the preamble, start frame delimiter, and, of course, the frame check sequence. When the transmitted frame arrives at the destination device, the receiving system calculates the CRC and compares the two. If they do not match, the station assumes an error has occurred and discards the frame.

The original standards for Ethernet defined the minimum frame size to be 64 octets and the maximum size to be 1,518 octets. This size includes everything from the Destination MAC Address field through the frame check sequence. In 1998, the IEEE issued 802.3ac, which extended the maximum frame size to 1,522 octets so that a VLAN tag could be included for evolving VLAN applications. The VLAN protocol enables an identifier or tag to be inserted in the frame so that the frame can be associated with the VLAN to which it pertains, so that logical traffic groups can be created for management purposes. IEEE 802.1Q defines the VLAN protocol.

If the four-byte VLAN tag is present in an Ethernet frame, it is found between the Source MAC Address and the Length/Type fields. The first two bytes of the tag comprise the 802.1Q tag type, always set to a value of 0×8100. This value indicates the presence of the VLAN tag and indicates to a receiving station that the normal Length/Type field will be found four bytes later in the frame.

The final two bytes of the VLAN tag are used for a variety of purposes. The first three bits represent a User Priority field used to assign a priority level to the information contained in the frame. The next bit is called a *Canonical Format Indicator* (CFI) and is used in Ethernet to indicate the presence of a *Routing Information Field* (RIF). The final 12 bits are the *VLAN Identifier* (VID) that identifies the VLAN that the frame belongs to. The format of an Ethernet frame with the VLAN identifier is shown in the following minitable:

Preamble (7 bytes)	Start Frame Delimiter (1 byte)	Dest. MAC Address (6 bytes)	Source MAC Address (6 bytes)	Length/ Type = 802.1Q Tag Type (2 bytes)	Tag Control Information (2 bytes)	Length/ Type (2 bytes)	MAC Client Data (0-n bytes)	Pad (0-p bytes)	Frame Check Sequence (4 bytes)

The *Interframe Gap* field defines a minimum idle period between the transmission of sequential frames. Sometimes called an *Interpacket Gap* field, it provides a recovery period between frames so that devices can prepare to receive the next frame. The minimum interframe gap is 9.6

microseconds for 10 Mbps Ethernet, 960 nanoseconds for 100 Mbps Ethernet, and 96 nanoseconds for 1 Gbps Ethernet.

In 1998, the IEEE approved 802.3ac, the standard that defines frame formats that support VLAN tagging on Ethernet networks.

The Special Case of Gigabit Ethernet

With the arrival in 1998 of the 802.3z standard for Gigabit Ethernet, a field was added to the end of the frame to ensure that the frame would be long enough to enable collisions to propagate to all stations on the network at high transmission speeds. The field is added as needed to bring the minimum length of the frame to 512 bytes. The Gigabit Ethernet Extension field is only required in half-duplex mode, since the collision protocol is not used in full-duplex mode.

Gigabit Ethernet Frame Bursting

When the 802.3z standard was released in 1998, the IEEE also added a *burst mode* option that enables a station to transmit a series of frames without relinquishing control of the transmission medium. It is only used with Gigabit and higher Ethernet speeds, and is only applicable in half-duplex mode. When a station is transmitting a series of short frames, this mode of operation can dramatically improve the overall performance of Gigabit Ethernet LANs.

The details of burst mode operation are interesting. Once a station has successfully transmitted a single frame, it has the option to continue to transmit additional frames until it reaches a burst limit of 8,192 byte times. An interframe gap, described earlier, is inserted between each frame. In this case, however, the transmitting station populates the interframe gap with extension bits, which are nondata symbols that serve as a keepalive message and serve to identify the actual data bits by the receiving station.

A Point of Contention: Jumbo Frames

In mid-1998, Alteon Networks put forward an initiative to increase the maximum size of the Data field from 1,500 octets to 9,000 octets. The initiative has been endorsed by a number of companies in the belief that larger frames would result in a more efficient network by reducing the number of frames that have to be processed. Furthermore, the 32-bit CRC used in traditional Ethernet loses its overall effectiveness in frames larger than 12,000 octets.

Alteon's proposal restricts the use of jumbo frames to full-duplex Ethernet links only, and it describes a link negotiation protocol that enables a station to determine if the station on the other end of the segment is capable of supporting jumbo frames before it chooses to transmit them.

Jumbo frames have been implemented in many systems, but they are not yet widespread. As demand grows and as IPv6 becomes more common, the technique will become more widely used.

Media Access Control (MAC) in Ethernet

Both half-duplex (one way at a time transmission) and full-duplex (two-way simultaneous) transmission schemes can be implemented in Ethernet networks. Both are discussed in the paragraphs that follow.

Half-Duplex Ethernet using CSMA/CD

In traditional CSMA/CD-based Ethernet, half-duplex transmissions are most commonly used. On LANs that rely on CSMA/CD, two or more stations share a common transmission medium. When a station wants to transmit a frame, it must wait for an idle period on the medium when no other station is transmitting and transmit the frame over the medium so that all other stations on the network hear the transmission. If another station tries to send data at the same time, a collision occurs. The transmitting station then sends a jamming signal to ensure that all stations are aware that the transmission failed due to a collision. The station then waits a random period of time before attempting to transmit again. This process is repeated until the frame is transmitted successfully.

The rules for transmitting a frame under command of CSMA/CD are simple and straightforward. The station monitors the network for the presence of an idle carrier or the presence of a transmitting station. This is the "carrier sense" part of CSMA/CD. If a transmission is detected, the transmitting station waits but continues to monitor for an idle period. If an active carrier is not immediately detected, and the period during which no carrier is equal to or greater than the interframe gap, the station begins frame transmission.

During the transmission, the frame monitors the medium for collisions. If it detects one, the station immediately stops transmitting and sends a

32-bit jamming sequence. If the collision is detected early in the frame transmission, the station completes the sending of the frame preamble before beginning the jamming sequence, which is transmitted to ensure that the length of the collision is sufficient to be noticed by the other transmitting stations.

After transmitting the jamming sequence, the station waits a random period of time before attempting to transmit again, a process known as *backoff*. By waiting a random period before transmitting, secondary collisions are reduced.

If multiple collisions occur, the transmission is still repeated, but the random delay time is progressively increased with each attempt. This serves to further reduce the probability of another collision. This process continues until one of the stations successfully transmits a frame without experiencing a collision. Once this has occurred, the station clears its collision counter.

The Backoff Process

As we discussed a moment ago, the backoff process is the process by which a transmitting station determines how long to wait after a collision before attempting to retransmit. Obviously, if all stations waited the same amount of time, secondary collisions would inevitably occur. Stations avoid this eventuality by generating a random number that determines the length of time they must wait before restarting the CSMA/CD process. This period is called the *backoff delay*.

The algorithm used to manage this process in Ethernet is officially known as *truncated binary exponential backoff*. We strongly urge you to memorize that name so that you can casually throw it out during cocktail parties—it's very impressive. When a collision occurs, each station generates a random number that always falls within a specific range of values. The random number determines the number of slot times it must wait before attempting to retransmit. The values increase by an exponential factor following each failed retransmission.

In the first attempt, the range is between 0 and 1. On the second attempt, it is between 0 and 3. For the third, it falls between 0 and 7. The process continues, expanding each time to a larger range of wait times. If multiple collisions occur, the range expands until it reaches 10 attempts, at which point the range is between 0 and 1,023, an enormous value. From that point on, the range of values remains fixed between 0 and 1,023. Practical limits, however, are established: If a station fails to transmit following

16 attempts, the MAC scheme issues an *excessive collision error*. The frame being transmitted is dropped, requiring that the application software reinitiate the transmission process.

Truncated binary exponential backoff is effective as a delay minimization mechanism when LAN traffic is relatively light. When the traffic is heavy, it is less effective. When traffic is excessively heavy, collisions cause excessive collision errors to be issued, an indication that traffic has increased to the point that a single Ethernet network can no longer effectively manage it. At that point, network administrators should step in and segment the network as part of a traffic management process.

Slot Time

Slot times are critical in half-duplex Ethernet networks. The time varies depending on the bandwidth of the LAN (the version of Ethernet deployed), but it is defined to be 512 bit times for networks operating at both 10 and 100 Mbps, and 4,096 bit times for Gigabit Ethernet. To guarantee the possibility of detecting collisions, the minimum transmission time for a complete frame must be at least the duration of one slot time, while the time required for collisions to propagate to all stations on the network must be less than a single slot period. Because of this design, a station cannot complete the transmission of a frame without detecting that a collision has occurred.

When stations transmit frames of data across the LAN, the transmission inevitably suffers unpredictable delays because of transit delays across variable length segments, delays resulting from the electronics through which they must pass, processing delays, and hub complexities. The length of time it takes for a signal to propagate between the two most separated stations on the LAN is called the maximum propagation delay. This time period is important because it helps network designers ensure that all stations can hear collisions when they occur and respond to them. In order for a station on the LAN to detect that a collision has occurred between the frame it is transmitting and that of another station, the signal must have the opportunity to propagate across the network so that another station can detect the collision and issue a jamming signal. That jamming signal must then make its way back across the network before being detected by the transmitting station. The sum of the maximum round-trip propagation delay and the time required to transmit the jamming signal define the length of the Ethernet slot time.

Slot times help to set limits on network size in terms of the maximum length of cable segments and the number of repeaters that can be in a given path. If the network becomes too large, *late collisions* can occur, a phenomenon that occurs when the evidence of a collision arrives too late to be useful to the MAC function. The transmitted frame is dropped and application software must detect its loss and initiate the retransmission sequence. As a result, if a collision occurs it will be detected within the first 512 bits (or 4,096 in the case of Gigabit Ethernet) of the transmission. This makes it easier for the Ethernet hardware to manage frame retransmissions following a collision. This is important because Ethernet hardware is relatively inexpensive, and because complexity adds cost, simple processes help to keep it inexpensive.

In Gigabit Ethernet environments, signals propagate a very short (and in fact unmanageable) distance within the traditional 512-bit timeframe. At gigabit transmission speeds, a 512-bit timeslot would support a maximum network diameter of about 20 meters. As a result, a carrier extension has been introduced to increase the timeslot to 4,096 bits. This enables Gigabit Ethernet to support networks as broad as 200 meters—plenty for certain applications in many metro environments.

Full-Duplex Ethernet

With the release of 802.3x, full-duplex Ethernet arrived on the scene, a protocol that operates effectively without the benefit of CSMA/CD. Full-duplex transmission enables stations to simultaneously transmit and receive data over a single point-to-point link. As a result, the overall throughput of the link is doubled.

For full duplex to operate on an Ethernet LAN, facilities must meet following stringent requirements:

- The medium itself must support simultaneous two-way transmission without interference. Media that meet this requirement include 10BaseT, 10BaseFL, 100BaseTX, 100BaseFX, 100BaseT2, 1000BaseCX, 1000BaseSX, 1000BaseLS, and 1000BaseT. 10Base5, 10Base2, 10BaseFP, 10BaseFB, and 100BaseT4 are incapable of supporting full-duplex transmission.

- Full-duplex operation is restricted to point-to-point connections between two stations. Because no contention exists (only two stations are available, after all), collisions cannot occur and CSMA/CD is not required. Frames can be transmitted at any time, limited only by the minimum interframe gap.

■ Both stations must be configured for full-duplex transmission.

Needless to say, full-duplex operation results in a far more efficient network since no collisions take place and therefore no bandwidth is wasted.

Pause Frames in a Full-Duplex Transmission

Full-duplex operation includes an optional flow control mechanism called a *pause frame*. These frames permit one station to temporarily flow-control traffic coming from the other station, with the exception of MAC frames used to manage transmission across the network. If one station transmits a high volume of frames resulting in serious congestion at the other end, the receiving station can ask the transmitter to temporarily cease transmission, giving it the opportunity to recover from the onslaught. Once the time period specified by the overloaded station expires, the transmitter returns to normal operation. Pause frames are bidirectional, meaning that either of the peer stations on the link can issue them.

Pause frames conform to the standard Ethernet frame format but also include a unique type field as well as a number of other variations on the traditional theme. For example, the destination address of the frame can be set to either the unique destination address of the station that the Pause command is directed toward or to a globally assigned multicast address that has been reserved by the IEEE for use in Pause frames. For those readers involved with bridging protocols, it is also reserved in the IEEE 802.1D bridging standard as an address type that will not be forwarded by bridges that it passes through, guaranteeing that the frame will not travel beyond its home segment.

The *Type* field is set to 88-08 in hex to indicate to a receiving device that the frame is a MAC frame.

The *MAC Control Opcode* field is set to 00-01 to indicate that the frame being used is a Pause frame. Incidentally, the PAUSE frame is the only type of MAC frame that is currently defined.

The *MAC Control Parameters* field contains a 16-bit value that specifies the length of the requested Pause in units of 512-bit times. Values can range from 00-00 to FF-FF in hexadecimal. If a second Pause frame arrives before the current Pause time has expired, its parameter will replace the current Pause time.

Finally, a 42-byte *Reserved* field (transmitted as all zeros) pads the Pause frame to the minimum Ethernet frame size.

The fields for a Pause frame are as follows:

Preamble (7 bytes)	Start Frame Delimiter (1 byte)	Dest. MAC Address (6 bytes) = (01-80-C2-00-00-01) or unique DA	Source MAC Address (6 bytes)	Length/ Type (2 bytes) = 802.3 MAC Control (88-08)	MAC Control Opcode (2 bytes) = Pause (00-01)	MAC Control Parameters (2 bytes) = (00-00 to FF-FF)	Reserved (42 bytes) = All zeros	Frame Check Sequence (4 bytes)

Link Aggregation

Link Aggregation, sometimes referred to as *trunking*, is an Ethernet feature used only in full-duplex mode. Essentially inverse multiplexing, link aggregation improves link performance and bandwidth availability between a pair of stations by enabling multiple physical facilities to serve as a single logical link. The link aggregation standard was created by the IEEE 802.3ad Working Group and entered into the formal standards in 2000.

Prior to the introduction of link aggregation, it was difficult to deploy multiple links between a pair of Ethernet stations because the Spanning Tree protocol commonly deployed in bridged Ethernet networks was designed to disable secondary paths between two points to prevent loops from occurring on the network. The link aggregation protocol eliminates this problem by enabling multiple links between two devices. They can be between two switches, between a switch and a server, or between a switch and an end user station, all offering the following advantages:

- Bandwidth can be increased incrementally. Prior to the introduction of link aggregation, bandwidth could only be increased by a factor of 10 by upgrading the 100 Mbps facility with a 1 Gbps facility.

- Link aggregation facilitates load balancing by distributing traffic across multiple facilities. Traffic can either be shared equally across the collection of aggregated facilities, or it can be segregated according to priority and transported accordingly.

- Multiple links imply a redundant transport capability. Under the link aggregation protocol, if a link in the collection fails, all traffic is simply redirected to another facility in the group, thus ensuring survivability.

- The link aggregation process is completely transparent to higher-layer protocols. It operates by inserting a "shim" protocol between the MAC protocol and the higher-layer protocols that reside above them. Each device in an aggregation of facilities transmits and receives using its own unique MAC address. As the frames reach the link aggregation protocol, the unique MAC addresses are shielded so that collectively they look like a single logical port to the higher layers.

Link aggregation is limited to point-to-point links and only works on links that operate in full-duplex mode. Furthermore, all the links in an aggregation group must operate at the same bandwidth level.

Up the Stack: The Ethernet Physical Layer

Over the years since Ethernet's arrival on the LAN scene, it has gone through a series of metamorphoses. Today, it still operates at a range of bandwidth levels and is found in a variety of applications ranging from the local area to limited long-haul deployments. In the section that follows, we discuss the various flavors of Ethernet, focusing on the physical layer. We begin with the venerable Base5.

10Base5

10Base5 was the first 10 Mbps Ethernet protocol. It operated over 10-millimeter coaxial cable and was commonly referred to as Thicknet because of the diameter of the cable. A word about nomenclature: 10Base5 represents a naming convention crafted during the creation of the 802 Committee standards. The 10 refers to 10 Mbps or the transmission speed, Base means that the transmission is digital baseband, and 5 refers to the 500-meter maximum segment length. Other names in the series are similar.

The original coax used in Thicknet installations was marked every 2.5 meters to indicate the point at which 10Base65 transceivers could be connected to the shared medium. They did not have to be connected at every 2.5-meter interval, but they *did* have to be placed at distances that were multiples of 2.5 meters. This separation served to minimize signal reflections that could degrade the quality of signals transmitted over the segment. Ten-millimeter Thicknet cable was typically bright yellow, with black indicator bands every 2.5 meters.

A Touch of Transylvania

In 10Base5 networks, transceivers were attached using a sort of clamp that wrapped around the cable and used needle-like attachments to pierce the insulation and make contact with the conductors within. These devices were often called "vampire taps" because of the way they punctured the shared cable to get at the "life blood" within. They were often referred to as nonintrusive taps because there was no requirement to interrupt network operations to install them.

In these early versions of Ethernet, end-user stations attached to the transceiver via a cable assembly known as an *attachment unit interface* (AUI). This AUI included the cable as well as a *network interface card* (NIC) that connected the cable to the NIC using a specialized 15-pin AUI connector. According to IEEE standards, this cable could be up to 50 meters in length, which means that stations could be that distance from the cable segment.

A 10Base5 coaxial cable segment could be up to 500 meters in length, and as many as 100 transceivers could be attached to a single segment, as long as they were separated by multiples of 2.5 meters. Furthermore, a 10Base5 segment could comprise a single piece of cable or could comprise multiple cable sections attached end to end, as long as impedance mismatches were managed properly to prevent signal reflections within the network.

Furthermore, multiple 10Base5 segments could be connected through segment repeaters, forming a large (but single and contiguous) collision domain. The repeaters regenerated the signal by transmitting it from one segment to another, amplifying the signal to maintain its signal strength throughout all the segments. The original standard permitted as many as five segments with four repeaters between any two stations. Three of the segments could be 1,500-meter coaxial segments, while the other two had to be point-to-point, 1,000-meter, interrepeater links. Each end of the three segments could have a 50-meter AUI cable, resulting in a maximum network length of 2,800 meters.

10Base2

The 10Base2 protocol supports 10 Mbps transmission over 5-millimeter coaxial cable, a configuration sometimes called *Thinnet*. It is also cheapnet and was the first standard introduced after traditional 10Base5 that relied on a different physical medium.

In many ways, 10Base2 resembles 10Base5. Both use 50-ohm coax, and both use a bus topology. At the physical layer, the two standards share common signal transmission standards, collision-detection techniques, and signal-encoding parameters. The primary advantage of 10Base2's thinner cable is that it is easier to install, manage, and maintain because it is lighter, more flexible, and less expensive than the thicker medium used by its predecessor. On the downside, the thinner cable offers greater transmission resistance and is therefore less desirable electrically. It supports a maximum segment length of 185 meters, compared to 10Base5's 500 meters, and no more than 30 stations per segment compared to 100 stations on 10Base5.

Although the spacing between stations on 10Base5 networks is critical for optimum performance, it is less so with 10Base2. The only limitation is that stations must be separated by no less than half a meter to minimize signal reflection.

Alas, vampires are extinct with the arrival of 10Base2, which uses a BNC T-connector instead of a vampire tap to attach transceivers to the cable segment. As Figure 3-2 shows, the vertical part of the T is a plug connector that attaches directly to the internal or external NIC, while the horizontal arm of the T comprises two sockets that can be attached to BNC connectors on each end of the cable segments that are to be connected. When a station is removed from the LAN, the T connector is replaced with a Barrel connector that provides a straight-through connection.

To further reduce signal reflection on the network and the resulting errors that can arise, each end of a 10Base2 segment must be terminated with a 50-ohm *Bayonet-Neill-Concelman* (BNC) terminator, which is electrically grounded for safety.

10Base2 Topologies   Two wiring schemes are supported by 10Base2. The T connectors described earlier support a daisy chain topology in which the cable segment loops from one computer to the next. Electrical terminators (50-ohm resistors) are installed on the unused connector at each end of the chained segment, as shown in Figure 3-3.

Figure 3-2
A BNC T-connector

Figure 3-3
A 50-ohm BNC terminator, used to terminate the end of an Ethernet segment

The alternative wiring scheme is a point-to-point model in which the segment connects a single station to a 10Base2 repeater. This wiring scheme is used in environments where the structure of the building makes it impractical or impossible to daisy chain multiple computers together. The segment is terminated at the computer station on one end and is connected to a repeater on the other end.

10Base2 Summary   10Base2 operates in half-duplex mode at 10 Mbps over a single 5-millimeter coaxial cable segment. Because the cable is thinner, it is less expensive and easier to install, but because it is a narrower cable than 10Base5, transmission distances over it are somewhat limited compared to its predecessor. Further, when the protocol is deployed over a daisy-chained topology, it can be somewhat difficult to administer and troubleshoot when problems occur. The maximum segment length in 10Base2 is 185 meters, and the maximum 30 stations per segment must be separated by at least one half meter.

10BaseT

The introduction of 10BaseT marked the arrival of modern Ethernet for business applications. It was the first version of the 802.3 standard to leave coax behind in favor of alternative media. 10BaseT provides a 10 Mbps transmission over two pair of Category Three twisted pair, sometimes called voicegrade twisted pair. Because it is designed to use twisted pair as its primary transmission medium, 10BaseT rapidly has become the most widely deployed version of Ethernet on the planet.

The 10BaseT protocol uses one pair to transmit data and the other pair to receive. Both pairs are bundled into a single cable that often includes two additional pairs that are not used in 10BaseT. This so-called multifunction cable may also include a bundle of fibers and one or more coaxial cables so that a single, large cable can transport voice, video, and data. Each end of the cable pairs used for 10BaseT is terminated with an eight-position RJ-45 connector.

Traditional 10BaseT connections are point to point. A 10BaseT cable, therefore, can have no more than two Ethernet transceivers, one at each end of the cable. In a typical installation, one end of the cable is usually connected to a 10BaseT hub, while the other end is to a computer's NIC or to an external 10BaseT transceiver.

In some cases, a pair of 10BaseT NICs is directly attached to each other using a special "crossover cable" that attaches the transmit pair of one station to the receive pair of the other, and vice versa. When attaching an NIC to a repeating hub, however, a normal cable is used and the crossover process is conducted inside the hub.

10BaseT Topology and Physical-Layer Issues When CAT 3 twisted pair is used, the maximum length per segment is 100 meters. Under some circumstances longer segments can be used as long as they meet signal quality transport standards. CAT 5 cable, which is higher quality than CAT 3 but is also more expensive, can extend the maximum segment length to 150 meters and still maintain quality stringency as long as installation requirements are met to ensure engineering compliance.

Although the original Ethernet products used a shared bus topology, modern Ethernet installations rely on a collection of point-to-point links that together make up a star-wired configuration, as shown in Figure 3-4. In this scheme, the point-to-point links connect to a shared central hub that is typically mounted in a central wiring closet to facilitate maintenance, troubleshooting, and provisioning.

Because 10BaseT uses separate transmit and receive pairs, it easily supports full-duplex transmission as an option. Needless to say, all components of the circuit—the media, the NIC, and the hub—must support full-duplex transport.

10BaseT Summary 10BaseT rapidly became the most widely deployed version of Ethernet LAN in the world because of its line speed and the fact that it operates over standard twisted-pair facilities. Also, because it supports a hubbed wiring scheme, it is easy to install, maintain, and troubleshoot. The only disadvantage it has is the fact that it supports shorter segment lengths than its predecessors, but its advantages far outweigh its disadvantages. In the functional domain of business, 10BaseT unquestionably dominates all other contenders.

▬▬ ▬▬ ▬▬ ▬▬

Figure 3-4
A star-wired Ethernet
LAN. Each station is
connected to a hub
(top) that is typically
(but not always) in a
wiring closet.

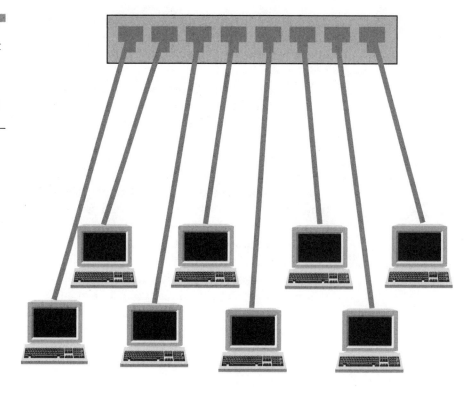

A Brief Aside: Nomenclature

With all this talk about wiring schemes, topologies, facilities, and logical and physical layouts, I thought it might be useful to take a brief side trip to explain two terms that are often misused and almost always misunderstood. These terms are *topology* and *wiring scheme*.

For the purposes of our discussion, topology refers to the *logical layout of the devices on the shared medium*. On the other hand, wiring scheme refers to the *physical layout of the network and the devices on it*. A simple example will clarify the difference.

Figure 3-5 shows a ring network—perhaps a token ring LAN, for example. Topologically, this network is a ring. However, to guard against ring failure caused by a physical breach, rings are often wired as shown in Figure 3-6. If Station A should become physically disconnected, for example, a relay in the hub closes and seals the breach in the ring. The wiring scheme of this LAN is a star, the form that best describes its physical layout. In fact, this layout is known as a *star-wired ring*. Similarly, the LAN shown in Figure 3-7 relies on a shared bus *topology*. However, because the entire shared

Figure 3-5
A token ring network

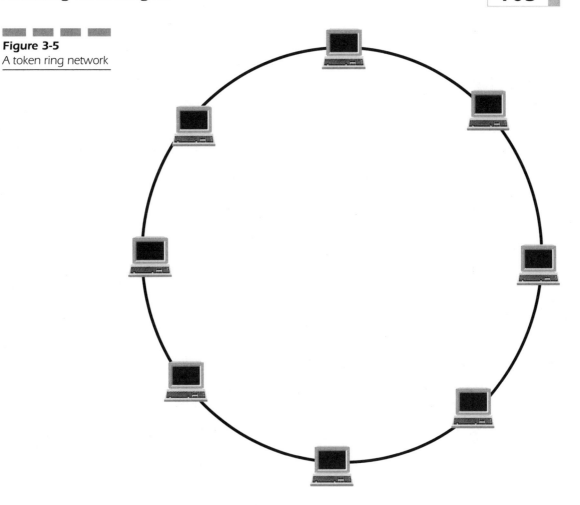

medium (the bus) is inside a physical hub, the single station LAN segments all emanate outward in a *star wiring scheme*.

10Broad36

Okay, if Ethernet works so well for LANs, perhaps it will work well for other applications like broadband video? 10Broad36 was designed to support the transmission of LAN traffic in addition to broadcast-quality video over the coax used in terrestrial *community antenna television* (CATV) systems. It provides 10 Mbps of bandwidth over a broadband cable system. The 36 in 10Broad36 refers to the maximum span length supported by the standard between two stations: 3,600 meters.

Figure 3-6
A star-wired LAN. Each station is connected to the central hub to improve the management and survivability of the network in the event of a breach of the medium.

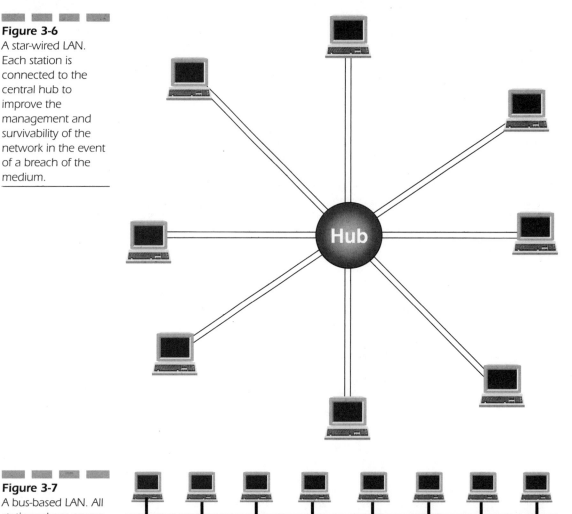

Figure 3-7
A bus-based LAN. All stations share access to the bus and therefore contend for its use.

CATV, which contrary to popular belief does *not* stand for *CA*ble *T*ele*Vi*sion but rather for *C*ommunity *A*ntenna *T*ele*V*ision, is an analog transmission scheme in which channels are created by dividing the available transmission spectrum into 6 MHz frequency bands and assigning each band to a broadcast channel. By managing the available bandwidth properly, broadband cable systems can support multiple services, including both data and video.

Broadband cable supports significantly longer segments than those supported by the original baseband 10Base5 and 10Base2. 10Broad36 segments can be as long as 1,800 meters, and all segments terminate at a head end (a signal origination point). As a result, a span of 3,600 meters can be accommodated (the far end of one segment through the head end to the far end of another segment). 10Broad36 transceivers attach directly to the cable both physically and electrically. The attachment cable can be as long as 50 meters, which adds another 100 meters to the overall end-to-end span of the system.

Fiber Optic Interrepeater Link (FOIRL)

As businesses went through their inevitable morphological shifts from monolithic designs to a more distributed functional model in response to the growing demands from customers in the metro marketplace, one of the demands that became somewhat urgent was the need for greater coverage distance without sacrificing performance or bandwidth availability. One of the first major evolutionary stages that emerged was FOIRL, a two-fiber design that delivers 10 Mbps of bandwidth. It was designed to provide greater distances over point-to-point connections between repeaters. The standard was originally released in 1987, considered an early standard by those who study modern metro networks, but it was soon updated in concert with changes in the technological demands placed on it. In 1993, 10BaseFL (fiber link) was released by the IEEE, which further refined and solidified the FOIRL concept.

FOIRL enables spans to be configured as long as 1,000 meters, significantly farther than any previous LAN standard. It should be noted that the newer 10BaseFL standard is backwards compatible with the older FOIRL technology. For example, a 10BaseFL transceiver can be deployed on one end of a facility with an older FOIRL transceiver used at the other. In this case, however, the maximum segment length is limited to the length specified by FOIRL (1,000 meters), not the 2,000-meter length supported by 10BaseFL.

10BaseF

The next member of the Ethernet pantheon is 10BaseF. Crafted as an enhancement to the earlier FOIRL standard, 10BaseF defines 10 Mbps operation over fiber.

10BaseF actually defines three segment types: 10BaseFL, 10BaseFB, and 10BaseFP. For reasons that will become clear in the paragraphs that follow, the three are not compatible with each other at the optical interface level.

10BaseFL 10BaseFL offers 10 Mbps transmission over a pair of fibers. The standard augments the FOIRL standard and supports a maximum segment length of 2,000 meters compared with FOIRL's 1,000 meters.

10BaseFL has a variety of applications. It can be used to connect computers, repeaters, or a computer to a repeater. All segments are point to point in 10BaseFL, with a single transceiver at each end of the segment. Typically, a computer attaches to the shared infrastructure via an external 10BaseFL transceiver. The NIC in the computer attaches to the external transceiver through a standard AUI cable. The transceiver, on the other hand, attaches to the two fibers using standard ST connectors. One fiber is used to transmit data, while the other is used to receive.

10BaseFL (fiber link, sometimes fiber loop) was originally designed to work with multimode fiber, because at the time of its creation multimode fiber was affordable and readily available, while single mode was only used by large service providers in long-haul installations due to cost.

The independent send and receive paths defined in 10BaseFL enable full-duplex mode to be supported as an optional configuration. In full-duplex mode, 10BaseFL supports 2,000-meter segment lengths. Furthermore, segment lengths are not restricted by the round-trip timing requirements of the CSMA/CD collision domain. In fact, segment lengths as great as 5 kilometers can be supported, and much longer segments can be installed if SMF is deployed.

High-quality multimode fiber and transceivers can support segment lengths of 5 kilometers. Even longer distances can be supported with the more expensive SMF.

10BaseFL is ideal for connecting between buildings. In addition to supporting longer segment lengths, fiber optic cables are immune to electrical hazards such as lightning strikes and ground currents that can occur when connecting separate buildings. Fiber is also immune to electrical noise that can be generated by motors or other electrical equipment.

10BaseFB 10BaseFB (fiber backbone) is primarily used to interconnect repeaters at 10 Mbps, using a unique synchronous signaling protocol. The signaling protocol enables the number of repeaters in a 10 Mbps Ethernet network to be extended. Like 10BaseFL, 10BaseFB segments can be as long as 2,000 meters.

Two factors limit the number of repeaters that can be deployed between two stations. First, each repeater adds a signal delay that can cause the time it takes for collisions to propagate throughout the network to exceed the maximum 512-bit time limit. Second, repeaters introduce random bit loss in the preamble field that can result in an overall reduction of the interframe gap below the required 9.6 microseconds (at 10 Mbps). 10BaseFB increases the number of allowable repeaters by reducing the amount of interframe gap loss. It does this by synchronizing the transmission interplay between the two repeaters. The interframe gap at the output of a normal repeater can be reduced by as many as eight bits. With 10BaseFB repeaters, the loss is reduced to two bits. This simplifies network design dramatically; suddenly, the only limiting factor on the size of the network is the time required for collisions to propagate.

10BaseFB can only be used on point-to-point links between repeaters, and both repeaters must support 10BaseFB. It *cannot* be used to connect a computer directly to a repeater. Furthermore, it does not support full-duplex mode.

10BaseFP   10BaseFP (fiber passive) relies on what is known as a "fiber optic passive star" topology. Based on passive optics, the passive star requires no power, can connect as many as 33 devices, and supports 500-meter segments. The star comprises a bundle of fibers that are fused so that a signal from the ingress fiber can be transmitted to a bundle of egress fibers at 10 Mbps. Although not widely used, this model is gaining acceptance in cable television infrastructures because it requires no local power for signal propagation. In effect, the star serves as a passive hub that receives signals from 10BaseFP transceivers and distributes them to all the other 10BaseFP transceivers. It supports half-duplex transmission only.

The Arrival of Fast Ethernet (100BaseT)

100BaseT refers to the specifications and media standards for 100 Mbps Ethernet, often called Fast Ethernet. Four standards have been defined: 100BaseTX, 100BaseFX, 100BaseT4, and 100BaseT2. Each will be discussed in this section.

All 100BaseT standards share a common MAC scheme, but each has its own unique physical-layer specification, sometimes referred to as a PHY, which usually defines the characteristics of the transceiver used in the system. The PHY can be integrated directly into a repeater or NIC, or attached externally.

100BaseTX

100BaseTX is a 100 Mbps Ethernet transmission scheme using two pairs of twisted pair, one for transmit, one for receive. A typical 100BaseTX wiring scheme uses a cable that often houses four pairs. However, because of the technology's high sensitivity to crosstalk, it is strongly recommended that the two remaining pairs be left unused. The standard supports 100-meter transmissions over 100-ohm CAT 5 UTP cable. CAT 5 is a higher-quality conductor than the CAT 3 cabling typically used with 10BaseT, and CAT 5 is rated for frequencies up to 100 MHz, while CAT 3 supports transmission up to 16 MHz.

All 100BaseTX segments are point-to-point with a single transceiver at each end of the cable. Most connections are designed to connect a computer to a hub; these hubs typically have an integrated transceiver in 100BaseTX installations.

100BaseFX

100BaseFX is a two-fiber architecture that also operates at 100 Mbps. The segment maximum length is 412 meters for half-duplex links and 2,000 meters (or longer) for full-duplex designs. 100BaseFX is basically the optical equivalent of the 100BaseTX standard described earlier; the cabling and connectors found in 100BaseTX installations are simply replaced with fiber and optical connectors. Both standards use the same 4B/5B signal-encoding scheme.

100BaseFX segments are point-to-point facilities with a single transceiver at each end of the link, and the transceivers are unique to the standard. 100BaseFX enables full-duplex mode, which allows the longer 2,000-meter segment design to be implemented. As before, longer distances can be accommodated with SMF.

100BaseT4

100BaseT4 supports 100 Mbps transmissions over 4 pairs of CAT 3 (or higher) twisted pair, a strong advantage given the low cost of CAT 3. In metro operations, where cost is a factor, such advantages tend to be noticed.

In 100BaseT4, one of the four pairs serves as a transmit pair, one as a receive pair, and two bidirectional pairs transmit or receive data. This model seems odd but actually makes a lot of sense. It ensures that one pair is always available for collision detection while the three remaining pairs are available for data transport. 100BaseT4 does not support the full-duplex mode of operation since it cannot support simultaneous transmitting and receiving at 100 Mbps.

100BaseT2

100BaseT2 is the only standard that supports 100 Mbps transmissions over two pairs of CAT 3 twisted pair. And unlike 100BaseTX, if additional pairs are available in the cable sheath, 100BaseT2 permits them to be used to transport voice or other LAN traffic. Even though 100BaseT2 is an innovative and powerful standard, it is not widely deployed today.

One Step Up: Gigabit Ethernet (1000BaseX)

1000BaseX, otherwise known as Gigabit Ethernet, is a relatively new arrival on the scene that has turned the LAN world on its ear and had an extraordinary impact on the metro world. Suddenly, corporate LANs are no longer hobbled by their glacially slow 10 Mbps networks or even their all-too-slow 100 Mbps LANs. As with the Fast Ethernet series, Gigabit Ethernet standards actually define a family of standards, described later.

Interestingly, these standards are based on physical-layer specifications adapted from the ANSI X3.230-1994 standard for Fibre Channel, a high-speed interface specification that has long been used for storage applications. By adopting the existing Fibre Channel specifications, 1000BaseX products found the market significantly easier to penetrate. It is important, however, that 1000BaseT does not use the Fibre Channel physical layer specifications and is not part of the 1000BaseX family of standards. 1000BaseT is described later in this part.

1000BaseLX

1000BaseLX is a fascinating standard, and the technologies that underlie it are equally interesting. The L in LX stands for long, because it relies on long wavelength lasers to transmit significant amounts of data over long-haul fiber. The lasers specified by the standard operate in the 1270- to 1355-nanometer range, and both single and multimode fiber are supported. Although long-wavelength lasers are more expensive than short-wavelength devices, they can transmit reliably over longer distances.

1000BaseSX

The antithesis of 1000BaseLX, 1000BaseSX (short) relies on short-wavelength lasers operating in the 770- to 860-nanometer range to transmit data over multimode fiber. These lasers are less expensive than their long-wavelength cousins and are ideal for short distances such as many of the fiber runs required in metro environments. The maximum segment length for these systems is between 275 and 550 meters, depending on the transmission mode (half- or full-duplex) and the type of fiber deployed.

1000BaseCX

The more things change, the more they stay the same. The C in 1000BaseCX stands for copper because it uses specially designed shielded balanced copper cables, sometimes called twinax. Segment distances are limited to 25 meters, but even with such a short run high-speed copper is clearly needed when interconnecting network components within a building.

1000Base-T

In June 1999, the IEEE released 802.3ab, the formal definition for 1000BaseT. The standard defines Gigabit Ethernet with transmission distances up to 100 meters over CAT 5 balanced copper facilities.

The 1000BaseT PHY relies on full-duplex baseband transmission over four pairs of CAT 5 cable. The aggregate data rate of 1000 Mbps is achieved by transmitting 250 Mbps over each pair. Hybrids and echo cancellation devices support full-duplex transmissions by allowing data to be transmitted and received on the same pairs simultaneously.

10-Gigabit Ethernet

To confirm the high-speed future of Ethernet, the IEEE 802.3ae Working Group has defined two physical-layer standards for 10 Gbps optical Ethernet. These include a LAN PHY and a WAN PHY. The LAN PHY does not differ from the existing LAN standard, while the WAN PHY is new. It specifies asynchronous 10 Gbps transmission using the traditional 802.3 Ethernet frame format, stipulates minimum and maximum frame sizes, and recommends a MAC scheme at the client interface. At 1550 nm, it will initially reach 40 kilometers, expanded to 80 kilometers in coming months.

Currently, 10 Gbps Ethernet is standardized for transport over fiber only; no copper twisted pair option is available as in Gigabit Ethernet. When the standard was first conceived, there was little thought given to a need for that kind of bandwidth to the desktop. However, the need is beginning to emerge and some chip manufacturers are now planning for 10 Gbps Ethernet over copper, using a design that will transmit 100 meters without a repeater. Working closely with the standards bodies, this capability will no doubt be reached soon.

Ethernet Summary

So why has Ethernet suddenly become such a powerful contender in the metro space? There are numerous reasons. As stated earlier, Ethernet has been around for a long time and is easy to use and trusted. Its advantages are scalability, rapid and easy provisioning, granular bandwidth management, and simplified network management due to its single protocol nature. Table 3-1 compares the relative characteristics of Gigabit Ethernet and the soon-to-be-released 10-Gigabit Ethernet.

Service Providers

A number of vendors have stepped up to the table and are now offering Ethernet access and transport over their networks. Table 3-2 displays a list of current Ethernet service providers, including their hardware vendors and offered services.

Table 3-1

Gigabit Ethernet versus 10 Gigabit Ethernet

Characteristic	Gigabit Ethernet	10 Gigabit Ethernet
Physical media	Single-mode and multimode fiber, WideWave Division Multiplexing (WWDM, 802.3z)	Single-mode and multimode fiber
Distance	3.11 miles	40 kilometers
MAC	Full, half duplex	Full duplex
Coding	8B/10B	64B/66B, SONET/SDH
Physical media dependencies	850 nm, 1300 nm, 1550 nm WWDM	850 nm, 1300 nm, 1550 nm
Supports 10 Gbps TDM and DWDM?	No	Yes
Packet size	64–1,514 octets	64–1,514 octets

Source: 10 Gbps Ethernet Alliance

Table 3-2

Current Ethernet service providers

Service provider	Hardware vendor	Services
Yipes	Extreme/Riverstone	Retail access, TLAN
Cogent	Chromatis/Cisco	Retail access
Telseon	Foundry/Riverstone	Retail and wholesale access
Metromedia	Cisco	Managed connectivity
Intellispace	Riverstone	Retail access
Looking Glass	Cisco	Retail and wholesale access
Everest Broadband	Appian	Retail access
Sigma Networks	CIENA/Cisco	Wholesale Ethernet transport
Louis Dreyfus Communications	Nortel/Sycamore	Retail access, TLAN
Enkido	Agnostic	Retail access
LDCOM Networks	Sycamore	European retail access
Netigy	Cisco	Retail access
Cox	Cisco/Riverstone	Retail access
Telia	Extreme/Riverstone	Wholesale access
BT	Cisco/Riverstone	Retail and wholesale access
Qwest	ONI	Retail and wholesale access

Source: Yankee Group

Alternative Premises Schemes

Of course, Ethernet is one of several premises schemes that drive traffic into the network core. Both alternative wired and wireless options exist; some of them are described here.

802.11

Another premises (some would say access) technology experiencing a great deal of attention today is 802.11, the set of standards that addresses *wireless LAN* (WLAN) considerations. IEEE 802.11 is a WLAN standard developed by the IEEE's 802 committee to specify an air interface between a wireless client and a base station, as well as among a variety of wireless clients. First discussed in 1990, the standard has evolved through six draft versions and won final approval on June 26, 1997.

802.11 Physical Layer

All 802 standards address themselves to both the physical and MAC layers. At the physical layer, IEEE 802.11 identifies three options for wireless LANs: diffused infrared, *Direct Sequence Spread Spectrum* (DSSS), and *Frequency Hopping Spread Spectrum* (FHSS).

Although the infrared PHY operates at a baseband level, the other two radio PHYs operate at 2.4 GHz, part of the *Industrial, Scientific, and Medical* (ISM) band. It can be used for operating WLAN devices and does not require an end-user license. All three PHYs specify support for 1 Mbps and 2 Mbps data rates.

Today 802.11 has evolved in a variety of directions, most of which have served to muddy the waters of the market. The original 802.11 WLAN standard was designed to operate at 2 Mbps within the 2.5 GHz range. Next came 802.11a, which operates at a whopping 54 Mbps in the 5 GHz range. The most recent addition is 802.11b, which operates at 11 Mbps in the 2.5 GHz range.

Careful readers will notice one significant disparity here: Even though 802.11a is a much faster transport standard, it operates in a frequency band that is incompatible with either of the other two, which are in fact compatible with each other. In Europe, 802.11a competes head to head with HiperLAN II, while in Japan it runs up against *High-Speed Wireless Access*

(HiSWAN). For most regions of the world, 802.11b seems to be the emerging de facto standard for wireless high-speed Ethernet. And because it is backwards compatible with the original 2.5 GHz standard, it is widely accepted as a next-generation solution.

802.11 MAC Layer

The 802.11 MAC layer, like CSMA/CD and token passing, presents the rules used to access the wireless medium. The primary services provided by the MAC layer are as follows:

- **Data transfer** Based on a *Carrier Sense Multiple Access with Collision Avoidance* (CSMA/CA) algorithm as the media access scheme.

- **Association** The establishment of wireless links between wireless clients and *access points* (APs).

- **Authentication** The process of conclusively verifying a client's identity prior to a wireless client associating with an AP. 802.11 devices operate under an open system where any wireless client can associate with any AP without verifying credentials. True authentication is possible with the use of the *Wired Equivalent Privacy Protocol* (WEP), which uses a shared key validation protocol similar to that used in *Public Key Infrastructures* (PKI). Only those devices with a valid shared key can be associated with an AP.

- **Privacy** By default, data is transferred "in the clear." Any 802.11-compliant device can potentially eavesdrop on PHY 802.11 traffic that is within range. WEP encrypts the data before it is transmitted using a 40-bit encryption algorithm known as RC4. The same shared key used in authentication is used to encrypt or decrypt the data; only clients with the correct shared key can decipher the data.

- **Power management** 802.11 defines an *active mode*, where a wireless client is powered at a level adequate to transmit and receive, and a *power save mode*, under which a client is not able to transmit or receive, but consumes less power while in a "standby mode" of sorts.

802.11 has garnered a great deal of attention in recent months, particularly with the perceived competition from Bluetooth, another short-distance wireless protocol. Significantly more activity is underway in the 802.11 space, however, with daily product announcements throughout the industry. Various subcommittees have been created that address everything from security to voice transport to QoS; it is a technology to watch.

Home Phoneline Networking Alliance (HomePNA)

The *Home Phoneline Networking Alliance* (HomePNA) is a not-for-profit association founded in 1998 by 11 companies to bring about the creation of uniform standards for the deployment of interoperable in-home networking solutions. The companies, 3Com, AMD, AT&T Wireless, Compaq, Conexant, Epigram, Hewlett-Packard, IBM, Intel, Lucent Technologies, and Tut Systems, are committed to the design of in-home communications systems that will enable the transport of LAN protocols across standard telephone inside wire.

HomePNA operates at 1 Mbps and enables the user to employ every RJ-11 jack in the house as both telephone and data ports for the interconnection of PCs, peripherals, and telephones. The standard enables for interoperability with existing access technologies, such as the *x-Type Digital Subscriber Line* (xDSL) and *Integrated Services Digital Network* (ISDN), and therefore provides a good migration strategy for *small office/home office* (SOHO) and telecommuter applications. The initial deployment utilizes Tut Systems' HomeRun product, which enables telephones and computers to share the same wiring and to simultaneously transmit. Extensive tests have been performed by the HomePNA, and although some minor noise problems caused by AC power noise and telephony-coupled impedance have resulted, they have been able to eliminate the effect through the use of low-pass filters between the devices causing the noise and the jack into which they are plugged.

Bluetooth

Bluetooth has been referred to as the *Personal Area Network* (PAN). It is a WLAN on a chip that operates in the unlicensed 2.4 GHz band at 768 Kbps, relatively slow compared to 802.11's 11 Mbps. It does, however, include a 56 Kbps backward channel and three voice channels, and it can operate at distances of up to 100 feet (although most pundits claim 25 feet for effective operation). According to a report from Allied Business Intelligence, the Bluetooth devices market will reach $2 billion by 2005, a nontrivial number.

The service model that Bluetooth supporters propose is one built around the concept of the mobile appliance. Bluetooth, named for a tenth-century Danish king (no, I don't know whether he had a blue tooth), is experiencing growing pains and significant competition for many good reasons from 802.11. Whether Bluetooth succeeds or not is a matter still open for discussion. It's far too early to tell.

Wireless Application Protocol (WAP)

Originally developed by Phone.com, the *Wireless Application Protocol* (WAP) has proven to be a disappointment for the most part. Because it is designed to work with *third-generation* (3G) wireless systems, and because 3G systems have not yet materialized in anticipated numbers, some have taken to defining WAP to mean "Wrong Approach to Portability." Germany's D2 network reports that the average WAP customer uses it less than two minutes per day, which is tough to make money on when service is billed on a usage basis. 3G will be the deciding factor; when it succeeds, WAP will succeed, unless 802.11's success continues to expand. Most believe it will.

The Mobile Appliance

The mobile appliance concept is enjoying a significant amount of attention of late because it promises to herald in a whole new way of using network and computer resources—*if it works as promised*. The problem with so many of these new technologies is that they overpromise and underdeliver, precisely the opposite of what they're supposed to do for a successful rollout. 3G, for example, has been billed as "the wireless Internet." Largely as a result of that billing, it has failed. It is *not* the Internet—far from it. The bandwidth isn't there, nor does it offer a device that can even begin to offer the kind of image quality that Internet users have become accustomed to. Furthermore, the number of screens that a user must go through to reach a desired site (I have heard estimates as high as 22!) is far too high. Therefore, until the user interface, content, and bandwidth challenges are met and satisfied, the technology will remain exactly that: a technology. There is no application yet, and *that's* what people are willing to pay money for.

Access Technologies

For the longest time, "access" described the manner in which customers reached the network for the transport of voice services. In the last 20 years, however, that definition has changed dramatically. In 1981, *20 years ago,* IBM changed the world when it introduced the PC, and in 1984 the Macintosh arrived, bringing well-designed and organized computing power to the proverbial masses. Shortly thereafter, hobbyists began to take advantage of emergent modem technology and created online databases, the first bulletin board systems that enabled people to send simple text messages to each other. This accelerated the modem market dramatically, and before long data became a common component of local loop traffic. At that time, there was no concept of Instant Messenger or of the degree to which e-mail would fundamentally change the way people communicate and do business. At the same time, the business world found more and more applications for data, and the need to move that data from place to place became a major contributor to the growth in data traffic on the world's telephone networks.

Today, then, *access* means much more than access to a network and its transport resources; access means *access to content and services*. It means putting into a place an access model that provides adequate bandwidth for an enterprise or residence customer to reach whatever online resources they need to reach, including voice, video, videoconferencing, storage farms, or databases. In the metro environment, it means providing access from an office location to a wide variety of bandwidth-intensive resources via the most cost-effective manner possible.

Marketplace Realities

According to a number of demographic studies that have recently been conducted, approximately 80 million households today host home office workers, and the number is growing rapidly. These numbers include both telecommuters and those who are self-employed and work out of their homes. They require the ability to connect to remote LANs and corporate databases, retrieve e-mail, access the Web, and in some cases conduct videoconferences with colleagues and customers. The traditional bandwidth-limited local loop is not capable of satisfying these requirements with traditional modem technology. Dedicated private-line service, which would solve the problem, is far too expensive as an option, and because it is dedicated it is not particularly efficient. Other solutions are required, and these have emerged in the form of access technologies that take advantage of either a conversion to end-to-end digital connectivity (ISDN) or expanded capabilities of the traditional analog local loop (*Digital Subscriber Line*, [DSL]). In some cases, a whole new architectural approach is causing excitement in the industry (*Wireless Local Loop* [WLL]). Finally, cable access has become a popular option as the cable infrastructure has evolved to a largely optical, all-digital system with high-bandwidth, two-way capabilities. We will discuss each of these options in the pages that follow.

ISDN

ISDN has been the proverbial technological roller coaster since its arrival as a concept in the late 1960s. Often described as the technology that took 15 years to become an overnight success, ISDN's level of success has been all over the map. Internationally, it has enjoyed significant uptake as a true, digital local loop technology. In the United States, however, because of competing and often incompatible hardware implementations, high cost, and spotty availability, its deployment has been erratic at best. In market areas where providers have made it available at reasonable prices, it has been quite successful. Furthermore, it is experiencing something of a renaissance today because of DSL's failure to capture the market. ISDN is tested, well understood, available, and fully functional. It can offer 128 Kbps of bandwidth *right now,* while DSL's capability to do so is less certain. ISDN has also seen its sales numbers climb in the last year because of the dramatic growth in demand for videoconferencing services. ISDN is ideally suited for this application because it enables a user to aggregate bandwidth as required in multiples of 64 Kbps. As videoconferencing use has grown, so too have ISDN's acceptance and deployment numbers.

ISDN Technology

The traditional local loop is analog. Voice traffic is carried from an analog telephone to the central office using a frequency-modulated carrier; once at the central office, the signal is digitized for transport within the wide area long-haul backbone. On the one hand, this is good because it means that the overall transmission path has a digital component. On the other hand, the loop is still analog, and as a result the true promise of an end-to-end digital circuit cannot be realized. The circuit is only as good as the weakest link in the chain, and the weakest link is clearly the analog local loop.

In ISDN implementations, local switch interfaces must be modified to support a digital local loop. Instead of using analog frequency modulation to represent voice or data traffic carried over the local loop, ISDN digitizes the traffic at the origination point, either in the voice set itself or in an adjunct device known as a *terminal adapter* (TA). The digital local loop then uses *Time Division Multiplexing* (TDM) to create multiple channels over which the digital information is transported, providing for a wide variety of truly integrated services.

The Basic Rate Interface (BRI)

ISDN has two implementations. The most common (and the one intended primarily for residence and small business applications) is called the *Basic Rate Interface* (BRI). In BRI, the two-wire local loop supports a pair of 64 Kbps digital channels known as B-Channels as well as a 16 Kbps D-Channel, which is primarily used for signaling but can also be used by the customer for low-speed (up to 9.6 Kbps) packet data. The B-Channels can be used for voice and data, and in some implementations they can be bonded together to create a single 128 Kbps channel for videoconferencing or other higher-bandwidth applications.

Figure 3-8 shows a typical ISDN BRI implementation, including the alphabetic reference points that identify the "regions" of the circuit and the generic devices that make up the BRI. In this diagram, the LE is the local exchange, or switch. The NT1 is the network termination device that serves as the demarcation point between the customer and the service provider. Among other things, it converts the two-wire local loop to a four-wire interface on the customer's premises. The *terminal equipment, type 1* (TE1) is an ISDN-capable device such as an ISDN telephone. This simply means that the phone is a digital device and is therefore capable of performing the voice digitization itself. A *terminal equipment, type 2* (TE2) is a non-ISDN-capable device, such as a *plain old telephone service* (POTS) telephone. In the event that a TE2 is used, a *terminal adapter* (TA) must be inserted between the TE2 and the NT1 to perform analog-to-digital conversion and rate adaptation.

Figure 3-8
The ISDN BRI

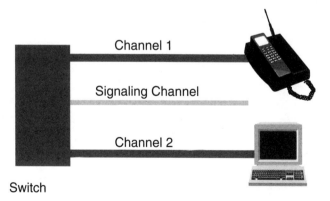

Channel 1

Signaling Channel

Channel 2

Switch

The reference points mentioned earlier identify circuit components between the functional devices just described. The U reference point is the local loop, the S/T reference point sits between the NT1 and the TEs, and the R reference point is found between the TA and the TE2.

BRI Applications

Although BRI does not offer the high bandwidth that other, more recent technologies such as DSL do, its multiple 64 Kbps channels provide reasonable transport capabilities for many applications. The three most common today are remote LAN access, videoconferencing (mentioned earlier), and Internet access. For the typical remote worker, the bandwidth available through BRI is more than adequate, and the ability to bond multiple 64 Kbps channels makes content transport a manageable task.

The Primary Rate Interface (PRI)

The other major implementation of ISDN is called the *Primary Rate Interface* (PRI). The PRI is really nothing more than the time-honored T-Carrier in that it is a four-wire local loop, uses *Alternate Mark Inversion* (AMI) and *Binary 8 Zero Substitution* (B8ZS) for ones-density control and signaling, and provides 24 channels of 64 Kbps that can be distributed among a collection of users as the customer sees fit (see Figure 3-9). In PRI, the signaling channel operates at 64 Kbps (unlike the 16 Kbps D-Channel in the BRI) and is not accessible by the user. It is used solely for signaling purposes; that is, it cannot carry user data. The primary reason for this is service protection; in the PRI, the D-Channel is used to control the 23 B-Channels and therefore requires significantly more bandwidth than the BRI D-Channel.

Figure 3-9
The ISDN PRI. The 23 (or 24) available channels can be combined and distributed however the user requires in multiples of 64 Kbps.

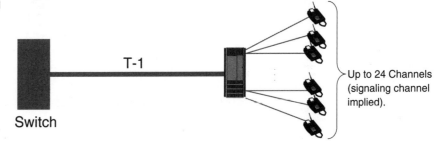

Switch

T-1

Up to 24 Channels (signaling channel implied).

Furthermore, the PRI standards enable multiple PRIs to share a single D-Channel, which makes the D-Channel's operational consistency all the more critical.

The functional devices and reference points are not appreciably different from those of the BRI. The local loop is still identified as the U reference point. In addition to an NT1, we now add an NT2, which is a service distribution device, usually a *private branch exchange* (PBX), which allocates the PRI's 24 channels to customers. This makes sense since PRIs are typically installed at businesses that employ PBXs for voice distribution. The S/T reference point is now divided; the S reference point sits between the NT2 and TEs, while the T reference point is found between the NT1 and the NT2.

PRI service also has the capability to provision B-Channels as super-rate channels to satisfy the bandwidth requirements of higher bit rate services. These are called H-Channels and are provisioned as shown here:

Channel	Bandwidth
H0	384 Kbps (6B)
H10	1.472 Mbps (23B)
H11	1.536 Mbps (24B)
H12	1.920 Mbps (30B)

PBX Applications

The PRI's marketplace is the business community, and its primary advantage is pair gain—that is, to conserve copper pairs by multiplexing the traffic from multiple user channels onto a shared, four-wire circuit. Inasmuch as a PRI can deliver the equivalent of 23 voice channels to a location over a single circuit, it is an ideal technology for a number of applications, including the interconnection of a PBX to a local switch, dynamic bandwidth allocation for higher-end videoconferencing applications, and interconnection between an *Internet service provider's* (ISP) network and that of the local telephone company.

Some PBXs are ISDN-capable on the line (customer) side, meaning that they have the capability to deliver ISDN services that emulate what would be provided over a direct connection to an ISDN-provisioned local loop. On the trunk (switch) side, the PBX is connected to the local switch via one or more T1s, which in turn provide access to the telephone network. This arrangement results in significant savings, faster call setup, more flexible administration of trunk resources, and the capability to offer a diversity of services through the granular allocation of bandwidth as required. As the use of *Internet Protocol* (IP) telephony in the metro environment continues to grow because of its capability to deliver unified messaging and other distinct advantages, ISDN PRI has emerged once again as a valuable ally because of its capability to provide a digital local loop on the line side of the PBX.

Videoconferencing

Although a single BRI provides adequate bandwidth for casual conferencing on a reduced-sized PC screen, PRI is preferable for high-quality, television-scale sessions. Its capability to provision bandwidth on demand makes it an ideal solution. All major videoconferencing equipment manufacturers have embraced PRI as an effective connectivity solution for their products and have consequently designed their products in accordance with accepted international standards, specifically H.261 and H.320.

ISDN is only one of the so-called twisted-pair solutions that extend the capacity and lifetime of the local loop. Another is DSL, a technology family that has achieved significant attention in the last couple of years. The attention is due to both successes and failings on the part of the technology. When it works, it works extremely well. When it doesn't, it tends to fail loudly and publicly.

Digital Subscriber Line (DSL)

The access technology that has enjoyed the greatest amount of attention in recent times is DSL. It provides a good solution for remote LAN access, Internet surfing, and access for telecommuters to corporate databases.

DSL came about largely as a direct result of the Internet's success. Prior to its arrival, the average telephone call lasted approximately four minutes, a number that central office personnel used while engineering switching systems to handle expected call volumes. This number was arrived at after nearly 125 years of experience designing networks for voice customers. They knew about Erlang theory, loading objectives, and peak-calling days/weeks/seasons, and they had decades of trended data to help them anticipate load problems.

In 1993, when the Internet—and with it, the World Wide Web—arrived, network performance became unpredictable as callers began to surf, often for hours on end. The average four-minute call became a thing of the past as online service providers such as *America Online* (AOL) began to offer flat-rate plans that did not penalize customers for long connect times. Service providers considered a number of solutions, including charging different rates for data calls than for voice calls, but none of these proved feasible until a technological solution was proposed. That solution was DSL.

It is commonly believed that the local loop is incapable of carrying more than the frequencies required to support the voice band. This is a misconception. ISDN, for example, requires significantly high bandwidth to support its digital traffic. When ISDN is deployed, the loop must be modified in certain ways to eliminate its designed bandwidth limitations. For example, some long loops are deployed with load coils that "tune" the loop to the voice band. They make the transmission of frequencies above the voice band impossible, but they enable the relatively low-frequency voice band components to be carried across a long local loop. High-frequency signals tend to deteriorate faster than low-frequency signal components, so the elimination of the high frequencies extends the transmission distance and reduces transmission impairments that would result from the uneven deterioration of a rich, multifrequency signal. These load coils therefore must be removed if digital services are to be deployed.

A local loop is only incapable of transporting high-frequency signal components if it is *designed* not to carry them. The capacity is still there; the network design, through load coil deployment, simply makes that additional bandwidth unavailable. DSL services, especially *Asymmetric DSL* (ADSL), take advantage of this "disguised" bandwidth.

DSL Technology

In spite of the name, DSL is an analog technology. The devices installed on each end of the circuit are sophisticated high-speed modems that rely on complex encoding schemes to achieve the high bit rates that DSL offers. Furthermore, several of the DSL services, specifically ADSL, G.Lite, *Very High-speed DSL* (VDSL), and *Rate Adaptive DSL* (RADSL), are designed to operate in conjunction with voice across the same local loop. ADSL is the most commonly deployed service and offers a great deal to both business and residence subscribers.

DSL Services

DSL comes in a variety of flavors designed to provide flexible, efficient, high-speed service across the existing telephony infrastructure. From a consumer point of view, DSL, especially ADSL, offers a remarkable leap forward in terms of available bandwidth for broadband access to the Web. As content has steadily moved away from being largely text-based and has become more graphical, the demand for faster delivery services has grown. DSL may provide the solution at a reasonable cost to both the service provider and the consumer.

It has also garnered the attention of the telephony service providers in a very different way. Because of its ubiquity and ease of use, cable modems have captured more than 70 percent of the broadband access market. The cable industry's ability to package high-speed access with desirable content and their ongoing willingness to offer IP telephony for $5 per line per month, or less, has gained them enormous marketshare and resulted in an overall downward trend in the number of revenue-generating access lines owned and operated by the incumbent telcos. A number of DSL manufacturers have entered into trials with telephone companies to demonstrate that DSL can be used as the physical layer transport scheme for broadband content—specifically, video on demand. They realize that one criterion for long-term success is the ability to deliver widely varied content, and DSL may be the solution that enables them to go head to head with the cable players.

Businesses will also benefit from DSL. Remote workers, for example, can rely on DSL for LAN and Internet access. Furthermore, DSL provides a good solution for VPN access as well as for ISPs looking to grow the bandwidth available to their customers. It is available in a variety of both symmetric and asymmetric services, and therefore offers a high-bandwidth access solution for a variety of applications. The most common DSL services are ADSL, *High-bit-rate DSL* (HDSL), HDSL-2, RADSL, *Symmetric High-Density DSL* (SHDSL), and VDSL. The special case of G.Lite, a form of ADSL, will also be discussed.

Asymmetric Digital Subscriber Line (ADSL)   When the World Wide Web and flat-rate access charges arrived, the typical consumer phone call went from roughly four minutes in duration to several times that. All the engineering that led to the overall design of the network based on an average four-minute hold time went out the window as the switches staggered under the added load. Never was the expression, "In its success lie the seeds of its own destruction," more true. When ADSL arrived, it provided the offload required to save the network.

The typical ADSL installation is shown in Figure 3-10. No change is required to the two-wire local loop, but minor equipment changes on either end, however, are required. First, the customer must have an ADSL modem at their premises. This device enables both the telephone service and a data access device, such as a PC, to be connected to the line.

The ADSL modem is more than a simple modem, in that it also provides the *Frequency Division Multiplexing* (FDM) process required to separate the voice and data traffic for transport across the loop. The device that actually does this, shown in Figure 3-11, is called a splitter, in that it splits the voice traffic away from the data. It is usually bundled as part of the ADSL modem, although it can also be installed as a card in the PC, as a stand-alone device at the demarcation point, or on each phone at the premises. The most common implementation is to integrate the splitter as part of the DSL modem; this, however, is the least desirable implementation because

Figure 3-10
Typical DSL architecture. Voice travels in the standard voiceband, while data travels in a higher-frequency channel.

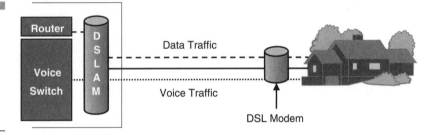

Figure 3-11
A DSL splitter

this design can lead to crosstalk between the voice and data circuitry inside the device. When voice traffic reaches the ADSL modem, it is immediately encoded in the traditional voice band and handed off to the local switch when it arrives at the central office. The modem is often referred to as an *ADSL Transmission Unit for Remote Use* (ATU-R). Similarly, the device in the central office is often called an ATU-C (for central office).

When a PC wants to transmit data across the local loop, the traffic is encoded in the higher-frequency band reserved for data traffic. The ADSL modem knows to do this because the traffic is arriving on a port reserved for data devices. Upon arrival at the central office, the data traffic does not travel to the local switch; instead, it stops at the ADSL modem that has been installed at the central office end of the circuit. In this case, the device is actually a bank of modems that serves a large number of subscribers and is known as a *Digital Subscriber Line Access Multiplexer* (DSLAM) (pronounced *dee-slam*). A DSLAM is shown in Figure 3-12.

Instead of traveling on to the local switch, the data traffic is now passed around the switch to a router, which in turn is connected to the Internet. This process is known as a *line-side redirect*.

The advantages of this architecture are fairly obvious. First, the redirect offloads the data traffic from the local switch so that it can go back to doing

Figure 3-12
A bay of DSLAMs
rack-mounted in a
central office

what it does best—switching voice traffic. Second, it creates a new line of business for the service provider. As a result of adding the router and connecting the router to the Internet, the service provider instantly becomes an ISP. This is a near-ideal combination, because it enables the service provider to become a *true service provider* by offering much more than simple access and transport.

As the name implies, ADSL provides two-wire asymmetric service; that is, the upstream bandwidth is different from the downstream. In the upstream direction, data rates vary from 16 to 640 Kbps, while the downstream bandwidth varies from 1.5 to 8 Mbps. Because most applications today are asymmetric in nature, this disparity poses no problem for the average consumer of the service.

A Word about the DSLAM   This device has received a significant amount of attention recently because of the central role that it plays in the deployment of broadband access services. Obviously, the DSLAM must interface with the local switch so that it can pass voice calls on to the PSTN. However, it often interfaces with a number of other devices as well. For example, on the customer side, the DSLAM may connect to a standard ATU-C, directly to a PC with a built-in *Network Interface Card* (NIC), to a variety of DSL services, or to an integrated access device of some kind. On the trunk side (facing the switch), the DSLAM may connect to IP routers as described before, to an ATM switch, or to some other broadband service provider. It therefore becomes the focal point for the provisioning of a wide variety of access methods and service types.

DSLAM: Centralized or Distributed?   One of the major challenges that service providers face as they roll out DSL is distance. The length of a DSL local loop has stringent distance limitations, and a rather high percentage of the loops in the United States exceed those limitations. Furthermore, a high percentage of would-be DSL customers are connected to the central office via a loop carrier system, a multiplexing arrangement that allocates 64 Kbps DS0 channels to each customer as a way to conserve physical facilities (pair gain). Unfortunately, it obviates the point of DSL by bandwidth-limiting each customer.

One reason these restrictions exist is because the DSLAM, the DSL service distribution device, resides in the central office. But what if it were placed closer to the customer, as shown in Figure 3-13? What if, for example, it were placed in the middle of a metro office park or in a neighborhood attached to the central office by an optical fiber? The result would be that all the limitations would disappear. Pair gain would cease to be a problem,

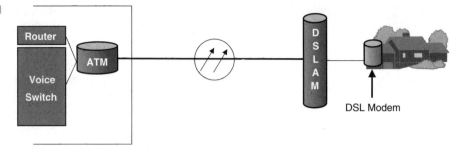

Figure 3-13
In this remote DSLAM network architecture, the DSLAM is located in the neighborhood or business park and is connected to the central office via an optical facility running ATM. The problem of distance and bandwidth is eliminated with this architecture, another example of core capability moving to the edge of the network.

because the fiber would deliver more than enough bandwidth to each customer. Furthermore, local loop length would no longer be a problem because the connection between the locally placed DSLAM and a customer location would rarely (if ever) exceed the maximum allowable distance.

This model of migrating the DSLAM from the central office to the periphery of the network is gaining widespread acceptance and is part of the convergence phenomenon that we have written extensively about. It is interesting that in order to achieve the benefits of convergence, *divergence* must be a part of the equation. By moving the intelligence and capability of the network closer to the end user, service is dramatically enhanced.

Symmetric High-Density DSL (SHDSL) A new flavor of DSL is just hitting the marketplace in the form of SHDSL. This standard is based on the preexisting standards that governed SDSL, HDSL, HDSL-2, *ISDN Digital Subscriber Line* (IDSL), and others, and therefore is easily implementable and well understood. It offers 30 percent longer reach than earlier versions of DSL, a fact that overcomes the single greatest barrier to DSL's widespread deployment. It also offers symmetric bandwidth over a 2-wire local loop at rates in excess of 2 Mbps, while a 4-wire version can offer bandwidth in the 5 Mbps range. The technology is rate adaptive and can therefore afford service providers the ability to offer a wide array of service plans. The primary market for SHDSL is the *small-to-medium* (SMB) business sector; it supports multiservice deployments including voice, video, data, and videoconferencing.

High Bit Rate DSL (HDSL) The greatest promise of HDSL is that it provides a mechanism for the deployment of four-wire T1 and E1 circuits without the need for span repeaters, which can add significantly to the cost of deploying data services. It also means that service can be deployed in a matter of days rather than weeks, something customers certainly applaud.

DSL technologies in general enable repeaterless facilities as long as 12,000 feet, while traditional 4-wire data circuits such as T1 and E1 require repeaters every 6,000 feet. Consequently, many telephone companies are now using HDSL behind the scenes as a way to deploy these traditional services. By the time a service provider factors in labor and hardware, repeater installations can cost upward of $100,000 to install, and if a loop is long, there may be multiples of them. Customers often do not realize that the T1 facility they plug their equipment into is being delivered using HDSL technology at the physical layer. The important thing is that they don't *need* to know. All the customer should have to care about is that a SmartJack is installed in the basement, and through that jack they have access to 1.544 Mbps or 2.048 Mbps of bandwidth—period. The real beneficiary of HDSL is the service provider because it provides pair gain. In the data world, where revenues per line are plummeting, every penny counts, especially on the capital side.

HDSL-2 HDSL-2 offers the same service that HDSL offers, with one added (and significant) advantage: It does so over a single pair of wire, rather than two. It also provides other advantages. First, it was designed to improve vendor interoperability by requiring less equipment at either end of the span (transceivers and repeaters). Second, it was designed to work within the confines of standard telephone company *Carrier Serving Area* (CSA) guidelines by offering a 12,000-foot wire-run capability that matches the requirements of CSA deployment strategies.

Rate-Adaptive DSL (RADSL) RADSL (pronounced 'rad-zel') is a variation of ADSL designed to accommodate changing line conditions that can affect the overall performance of the circuit. Like ADSL, it relies on DMT encoding, which selectively "populates" subcarriers with transported data, thus allowing for granular rate setting.

Very High-Speed DSL (VDSL) VDSL shows promise as a provider of extremely high access bandwidth, as much as 52 Mbps over a short local loop. VDSL requires *Fiber-to-the-Curb* (FTTC) architecture and recommends ATM as a switching protocol; from a fiber hub, copper tail circuits deliver the signal to the business or residential premises. Bandwidth available through VDSL ranges from 1.5 to 6 Mbps on the upstream side, and from 13 to 52 Mbps on the downstream side. Obviously, the service is distance-sensitive and actual achievable bandwidth drops as a function of distance. Nevertheless, even a short loop is respectable when such high bandwidth levels can be achieved. With VDSL, 52 Mbps can be reached over a loop length of up to 1,000 feet, a not unreasonable distance by any means.

G.Lite   Because the installation of splitters had proven to be a contentious and problematic issue, the need arose for a version of ADSL that did not require them. That version became known as either ADSL Lite or G.lite (after the ITU-T G-Series standards that govern much of the ADSL technology). In 1997, Microsoft, Compaq, and Intel created the *Universal ADSL Working Group* (UAWG)[2], an organization that grew to nearly 50 members dedicated to the task of simplifying the rollout of ADSL. In effect, the organization had four stated goals:

- To ensure that analog telephone service will work over the G.Lite deployment without remote splitters, in spite of the fact that the quality of the voice may suffer slightly due to the potential for impedance mismatch.
- To maximize the length of deployed local loops by limiting the maximum bandwidth provided. Research indicates that customers are far more likely to notice a performance improvement when migrating from 64 Kbps to 1.5 Mbps than when going from 1.5 Mbps to higher speeds. Perception is clearly important in the marketplace, so the UAWG chose 1.5 Mbps as their downstream speed.
- To simplify the installation and use of ADSL technology by making the process as "plug and play" as possible.
- To reduce the cost of the service to a perceived reasonable level.

Of course, G.lite is not without its detractors. A number of vendors have pointed out that if G.lite requires the installation of microfilters at the premises on a regular basis, then true splitterless DSL is a myth, since microfilters are in effect a form of splitter. They contend that if the filters are required anyway, then they might as well be used in full-service ADSL deployments to guarantee high-quality service delivery. Unfortunately, this

---

[2]The group "self-dissolved" in the summer of 1999 after completing what they believed their charter to be.

flies in the face of one of the key tenets of G.lite, which is to simplify and reduce the cost of DSL deployment by eliminating the need for an installation dispatch (a "truck roll" in the industry's parlance).

The key to G.lite's success in the eyes of the implementers has been to eliminate the dispatch, minimize the impact on traditional POTS telephones, reduce costs, and extend the achievable drop length. Unfortunately, customers still have to be burdened with the installation of microfilters, and coupled noise on POTS is higher than expected. Many vendors argue that these problems largely disappear with full-feature ADSL using splitters; a truck dispatch is still required, but again, it is often required to install the microfilters anyway, so no net loss occurs. Furthermore, a number of major semiconductor manufacturers support both G.lite and ADSL on the same chipset, so the decision to migrate from one to the other is a simple one that does not necessarily involve a major replacement of internal electronics.

Service Provider Issues

The greatest challenges facing those companies looking to deploy DSL are competition from cable and wireless, pent-up customer demand, installation issues, and plant quality.

Competition from cable and wireless companies represents a threat to wireline service providers on several fronts. As discussed earlier, cable modems enjoy a significant amount of press and are therefore gaining well-deserved marketshare. The service they provide is for the most part well received and offers high-quality, high-speed access. Wireless, on the other hand, is a slumbering beast. It has largely been ignored as a serious competitor for data transport, but recent announcements prove that its viability as a contender is real. 3G was the original "golden child technology," but its popularity has slipped as perception of its capability to make good on promises of services has gone decidedly downhill. GSM and GPRS are powerful contenders, but only have a minor share of the market in the United States, although that marketshare is growing.

The second challenge is unmet customer demand. If DSL is to satisfy the broadband access requirements of the marketplace, it must be made available throughout *Incumbent Local Exchange Carrier* (ILEC) service areas. This means that incumbent providers must equip their central offices with DSLAMs that will provide the line-side redirect required as the initial stage of DSL deployment. The law of primacy is evident here: The ILECs must get to market first with broadband offerings if they are to achieve and keep a place in the burgeoning broadband access marketplace. Unfortunately, a number of barriers, most notably regulatory in nature, have made short shift of their plans for widespread DSL deployment.

The Regulatory Conundrum

As already discussed in Part 2, the greatest challenge that service providers have faced is the requirement to provide true universal service in rural areas as well as in metropolitan and suburban areas. Telephone companies began being subsidized in the 1930s under Roosevelt to provide universal service, and the majority of the local loops in the United States are subsidized to the tune of anywhere from $3 to $15 per month. As a result, the incumbent service providers often charge significantly less for their service than it actually costs them to provide it.

Only the ILECs enjoy these subsidies, which means that CLECs are heavily penalized right out of the starting gate. As a result, they typically ignore the residence market in favor of the far more lucrative business markets, particularly the metro space. Business markets are significantly easier and less costly to provision than residence installations because of the dominance of *multitenant buildings* (MTUs).

New technologies always convey a temporary advantage to the initial innovator. Innovation involves a significant risk, yet the innovator usually makes a profit as a reward for taking the initial risk, thus allowing him or her to recover the cost of innovation. As technology advances, the innovation is inevitably surpassed by another one, so the advantage is fleeting. The relationship, however, fosters a zeal for ongoing entrepreneurial development.

Under the regulatory rule set that exists in many countries today, incumbent providers are required to open their networks through element unbundling and to sell their resources, including new technologies, at a wholesale price to their competitors. In the minds of regulators, this creates a competitive marketplace, but it is artificial. In fact, the opposite happens: The incumbents lose their incentive to invest in new technology, and innovation stalls. Under the wholesale unbundling requirements, the rewards for innovative behavior are socialized, while the risks undertaken by the incumbents are privatized. Why should they invest and take substantial economic risk when they are required by law to immediately share the rewards with their competitors?

The Powell *Federal Communications Commission* (FCC) seems to understand this issue and has taken a number of steps to correct it. A number of studies indicate that the widespread deployment of broadband access in the

United States would inject as much as $400 billion into the national economy, a number that tends to get the attention of legislators and regulators. At the time of this writing, a decision is pending to reclassify DSL and cable modem technologies as information services rather than telecom services. This would eliminate them from unbundling requirements and perhaps provide incentives to the incumbents to accelerate investment in the broadband local loop.

The third and fourth challenges to rapid and ubiquitous DSL deployment are installation issues and plant quality. A significant number of impairments have proven to be rather vexing for would-be deployers of widespread DSL. These challenges fall into two categories: electrical disturbances and physical impairments.

Electrical Disturbances

The primary cause of electrical disturbance in DSL is crosstalk, caused when the electrical energy carried on one pair of wires "bleeds" over to another pair and causes interference (noise) there. Crosstalk exists in several flavors. *Near-end crosstalk* (NEXT) occurs when the transmitter at one end of the link interferes with the signal received by the receiver at the same end of the link. *Far-end crosstalk* (FEXT) occurs when the transmitter at one end of the circuit causes problems for the signal picked up by a receiver at the far end of the circuit. Similarly, problems can occur when multiple DSL services of the same type exist in the same cable and interfere with one another. This is referred to as self-NEXT or self-FEXT. When different flavors of DSL interfere with one another, the phenomenon is called Foreign-NEXT or foreign-FEXT. Other problems can cause errors in DSL and therefore a limitation in the maximum achievable bandwidth of the system. Simple *radio frequency interference* (RFI) can find its way into the system, and impulse, Gaussian, and random noise that exists in the background can affect signal quality even at extremely low levels.

Physical Impairments

The physical impairments that can have an impact on the performance of a newly deployed DSL circuit tend to be characteristics of the voice network that typically have a minimal effect on simple voice and low-speed data services. These include load coils, bridged taps, splices, mixed gauge loops, and weather conditions.

Load Coils and Bridged Taps

We have already mentioned the problems that load coils can cause for digital transmission. Because they limit the frequency range that is allowable across a local loop, they can seriously impair transmission. Bridged taps are equally problematic. When a multipair telephone cable is installed in a newly built-up area, it is generally some time before the assignment of each pair to a home or business is actually made. To simplify the process of installation when the time comes, the cable's wire pairs are periodically terminated in terminal boxes installed every block or so (sometimes called B-Boxes; see Figure 3-14). Each pair may make multiple appearances on a city street, waiting for assignment to a customer. When the time comes to assign a particular pair, the installation technician will simply go to the terminal box where that customer's drop appears and cross-connect the loop to the appearance of the cable pair (a set of lugs) to which that customer has been assigned. This eliminates the need for the time-consuming process of splicing the customer directly into the actual cable.

Figure 3-14
Terminal box, sometimes known as a B-Box

Unfortunately, this efficiency process also creates a problem. Although the customer's service has been installed in record time, unterminated appearances of the customer's cable pair can be found in each of the terminal boxes along the street. These so-called bridged taps present no problem for analog voice, but they can be a catastrophic source of noise due to signal reflections that occur at the copper-air interface of each bridged tap. If DSL is to be deployed over the loop, the bridged taps must be removed. Although the specifications indicate that bridged taps do not cause problems for DSL, the actual deployment says otherwise. To achieve the bandwidth that DSL promises, the taps must be terminated.

Splices and Gauge Changes

Splices can result in service impairments due to cold solder joints, corrosion, or weakening effects caused by the repeated bending of wind-driven aerial cable. Gauge changes tend to have the same effects that plague circuits with unterminated bridged taps: When a signal traveling down a wire of one gauge jumps to a piece of wire of another gauge, the signal is reflected, resulting in an impairment known as intersymbol interference. The use of multiple gauges is common in a loop deployment strategy because they enable outside plant engineers to use lower-cost, small gauge wire where appropriate, cross-connecting it to larger-gauge, lower-resistance wire where necessary.

Weather

Weather is perhaps a bit of a misnomer; the real issue is moisture. One of the greatest challenges facing DSL deployment is the age of the outside plant. Much of the older distribution cable uses inadequate insulation between the conductors (paper in some cases). Occasionally, the outer sheath cracks, allowing moisture to seep into the cable itself. The moisture causes crosstalk between wire pairs that can last until the water evaporates. Unfortunately, this can take a considerable amount of time and result in extended outages.

Solutions

All these factors have solutions. The real question is whether they can be remedied at a reasonable cost. Given the growing demand for broadband access, there seems to be little doubt that the elimination of these factors would be worthwhile at all reasonable costs, particularly considering how competitive the market for the local loop customer has become.

The electrical effects, largely caused by various forms of crosstalk, can be reduced or eliminated in a variety of ways. Physical cable deployment standards are already in place that, when followed, help to control the amount of near and far-end crosstalk that can occur within binder groups in a given cable. Furthermore, filters have been designed that eliminate the background noise that can creep into a DSL circuit.

Physical impairments can be controlled to a point, although to a certain extent the service provider is at the mercy of their installed network plant. Obviously, older cable will present more physical impairments than newer cable, but the service provider can take certain steps to maximize their success rate when entering the DSL marketplace. The first step that they can take is to prequalify local loops for DSL service to the greatest extent possible. This means running a series of tests using *Mechanized Loop Testing* (MLT) to determine whether each loop's transmission performance falls within the bounds established by the service provider as well as the existing standards and industry support organizations.

Cable-Based Access Technologies

In 1950, Ed Parsons placed an antenna on a hillside above his home in Washington State, attached it to a coaxial cable distribution network, and began to offer television service to his friends and neighbors. Prior to his efforts, the residents of his town were unable to pick up broadcast channels because of the blocking effects of the surrounding mountains. Thanks to Parsons, CATV was born; from its roots came cable television.

Since that time the cable industry has grown into a powerhouse. In the United States alone, 13,000 head ends deliver content to 70 million homes in more than 22,000 communities with over more than a million miles of coaxial and fiber-optic cable. As the industry's network has grown, so too have the aspirations of those deploying it. Their goal is to make it much more than a one-way medium for the delivery of television and pay per view; they want to provide a broad spectrum of interactive, two-way services that will enable them to compete head-to-head with the telephony industry. To a large degree, they are succeeding. The challenges they face, however, are daunting.

Playing in the Broadband Game

Unlike the telephone industry that began its colorful life under the scrutiny of a small number of like-minded individuals (such as Alexander Graham Bell and Theodore Vail, among others), the cable industry came about thanks to the combined efforts of hundreds of innovators, each building on Parson's original concept. As a consequence, the industry, while enormous, is in many ways fragmented. Powerful industry leaders like John Malone and Gerald Levine were able to exert enormous power to unite the many companies, turning a loosely cobbled-together collection of players into cohesive, powerful corporations with a shared vision of what they were capable of accomplishing.

Today, the cable industry is a force to be reckoned with, and upgrades to the original network are underway. This is a crucial activity that will ensure the success of the industry's ambitious business plan and provide a competitive balance for the traditional telcos.

The Cable Network

The traditional cable network is an analog system based on a tree-like architecture. The head end, which serves as the signal origination point, serves as the signal aggregation facility. It collects programming information from a variety of sources, including satellite and terrestrial feeds. Head-end facilities often look like a mushroom farm; they are typically surrounded by a variety of satellite dishes (see Figures 3-15 and 3-16).

Figure 3-15
Satellite receive antennas at a cable head-end facility

Figure 3-16
A Very Small Aperture
Terminal (VSAT) dish
on a red clay tile roof
in a Spanish pueblo

The head end is connected to the downstream distribution network by one-inch-diameter rigid coaxial cable, as shown in Figure 3-17. That cable delivers the signal, usually a 450 MHz collection of 6 MHz channels, to a neighborhood, where splitters divide the signal and send it down half-inch-diameter semirigid coax that typically runs down a residential street. At each house, another splitter (see Figure 3-18) pulls off the signal and feeds it to the set-top box in the house over the drop wire, a "local loop" of flexible quarter-inch coaxial cable.

Although this architecture is perfectly adequate for the delivery of one-way television signals, its shortcomings for other services should be fairly obvious to the reader. First of all, it is, by design, a broadcast system. It does not typically have the capability to support upstream traffic (from the customer toward the head end) and is therefore not suited for interactive applications. Second, because of its design, the network is prone to significant failures that have the potential to affect large numbers of customers. The tree structure, for example, means that if a failure occurs along any "branch" in the tree, every customer from that point downward loses service. Contrast this with the telephone network where customers have a dedicated local loop over which their service is delivered. Second, because the system is analog, it relies on amplifiers to keep the signal strong as it is propagated downstream. These amplifiers are powered locally; they do not have access to central office power, as devices in the telephone network do.

Figure 3-17
The layout of the traditional cable distribution network. Modern cable networks replace the ones with optical fiber, often deployed in a ring architecture for survivability.

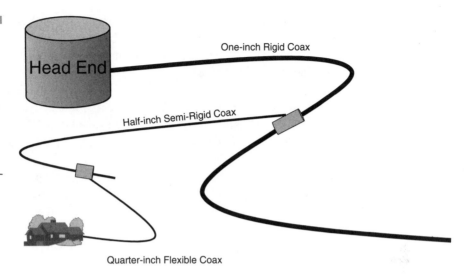

One-inch Rigid Coax

Half-inch Semi-Rigid Coax

Quarter-inch Flexible Coax

Figure 3-18
A signal splitter in a residential cable installation. The splitter pulls the signal off the distribution and feeds it to each residence.

Consequently, a local power failure can bring down the network's capability to distribute service in that area.

The third issue is one of customer perception. For any number of reasons, the cable network is generally perceived to be not as capable or as reliable as the telephone network. As a consequence of this perception, the cable industry is faced with the daunting challenge of convincing potential voice and data customers that they are in fact capable of delivering high-quality service. Some of the concerns are justified. In the first place, the telephone network has been in existence for almost 125 years, during which time its operators have learned how to optimally design, manage, and operate it in order to provide the best possible service. The cable industry, on the other hand, came about 50 years ago and didn't benefit from the rigorously administered, centralized management philosophy that characterized the telephone industry. Additionally, the typical 450 MHz cable system did not have adequate bandwidth to support the bidirectional transport requirements of new services.

Furthermore, the architecture of the legacy cable network, with its distributed power delivery and tree-like distribution design, does not lend itself to the same high degree of redundancy and survivability that the telephone network offers. Consequently, cable providers have been hard-pressed to convert customers who are vigorously protective of their telecommunications services.

Evolving Cable Systems

Faced with these harsh realities and the realization that the existing cable plant could not compete with the telephone network in its original analog incarnation, cable engineers began a major reworking of the network in the early 1990s. Beginning with the head end and working their way outward, they progressively redesigned the network to the extent that in many areas of the country their coaxial "local loop" was capable of competing on equal footing with the telco's twisted pair, and in some cases beating it.

The process they have used in their evolution consists of four phases. In the first phase, they converted the head end from analog to digital. This allowed them to digitally compress the content, resulting in far more efficient utilization of the available bandwidth. Second, they undertook an ambitious physical upgrade of the coaxial plant, replacing the one-inch trunk and half-inch distribution cable with optical fiber. This brought about several desirable results. First, by using a fiber feeder, network designers were able to eliminate a significant number of the amplifiers responsible for the failures the network experienced due to power problems in the field. Second, the fiber makes it possible to provision significantly more bandwidth than coaxial systems allow. Third, because the system is digital, it suffers less from noise-related errors than its analog predecessor did. Finally, an upstream return channel was provisioned, as shown in Figure 3-19, which makes the delivery of true interactive services possible, such as voice, Web surfing, and videoconferencing.

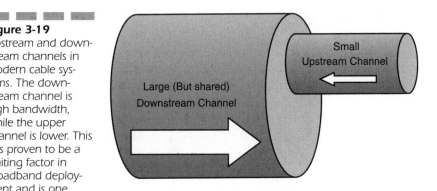

Figure 3-19
Upstream and down-
stream channels in
modern cable sys-
tems. The down-
stream channel is
high bandwidth,
while the upper
channel is lower. This
has proven to be a
limiting factor in
broadband deploy-
ment and is one
reason why DSL is
favored over cable
modems.

The third phase of the conversion had to do with the equipment provisioned at the user's premises. The analog set-top box has now been replaced with a digital device that can take advantage of the capabilities of the network, including access to the upstream channel. It decompresses digital content, performs content ("stuff") separation, and provides the network interface point for data and voice devices.

The final phase is business conversion. Cable providers look forward to the day when their networks will compete on equal footing with the twisted-pair networks of the telephone company, and customers will see them as viable competitors. In order for this to happen, they must demonstrate that their network is capable of delivering a broad variety of competitive services, that the network is robust, that they have *operations support systems* (OSSs) that will guarantee the robustness of the infrastructure, and that they are cost competitive with incumbent providers. They must also create a presence for themselves in the business centers of the world. Today they are almost exclusively residential service providers; if they are to break into the potentially lucrative business market, they must have a presence there.

Cable Modems

As cable providers have progressively upgraded their networks to include more fiber in the backbone, their plan to offer two-way access to the Internet has become a reality. Cable modems offer access speeds of up to 10 Mbps, and so far the market uptake has been spectacular.

Cable modems provide an affordable option to achieve high-speed access to the Web, with current monthly subscription rates in the neighborhood of $40. They offer asymmetric access, that is, a much higher downstream speed than upstream, but for the majority of users this does not represent a problem since the bulk of their use will be for Web surfing, during which most the traffic travels in the downstream direction anyway.

Although cable modems do speed up access to the Web and other online services by several orders of magnitude, a number of downsides must be considered. The greatest concern that has been voiced about cable modems is security. Because cable modems are "always on," they represent an easy entry point for hackers looking to break into machines. It is therefore critical that cable subscribers use some form of firewall software or a router that has the capability to perform filtering.

Cable Access in the Metro Arena

Cable is virtually 100 percent dedicated to the residence market. It has achieved minimal penetration in the business world for a variety of reasons, not the least of which were concerns over service quality, bandwidth, and sustainable availability. For the most part, these concerns have been overcome; cable companies have hired hundreds of former telco employees to help them develop centralized *Operations, Administration, Maintenance, and Provisioning* (OAM&P) procedures to facilitate peer-to-peer competition with the incumbent telecom players. They face an uphill battle in the metro enterprise world, however; their presence there is minimal, as is their ability to meet the growing demands of that particular customer base. So although they may become contenders in the future, today their ability to compete for metro business is small.

Wireless Access Technologies

It is only in the last few years that wireless access technologies have advanced to the point that they are being taken seriously as contenders for the broadband local loop market. Traditionally, a minimal infrastructure was put in place, and it was largely bandwidth-bound and error-prone to the point that wireless solutions were not considered serious contenders.

Wireless Access

To understand wireless communications, it is necessary to examine both radio and telephone technologies, because the two are inextricably intertwined. In 1876, Alexander Graham Bell, a part-time inventor and a teacher of hearing-impaired students, invented the telephone while attempting to resolve the challenge of transmitting multiple telegraph signals over a shared pair of wires. His invention changed the world forever.

In 1896, a mere 20 years later, Italian engineer and inventor Guglielmo Marconi developed the spark gap radio transmitter, which eventually enabled him to transmit long-wave radio signals across the Atlantic Ocean as early as 1901. Like Bell, his invention changed the world; for his contributions, he was awarded the Nobel Prize in 1909.

It wasn't until the 1920s, though, when these two technologies began to dovetail, that their true promise was realized. Telephony provided interpersonal, two-way, high-quality voice communications, but it required the user to be stationary. Radio, on the other hand, provided mobile communications but was limited by distance, environmentally induced signal degradation, and spectrum availability. Whereas telephony was advertised as a universally available service, radio was more of a catch-as-catch-can offering that was subject to severe blocking. If a system could be developed that combined the signal quality and ubiquity of telephony with the mobility of radio, however, a truly promising new service offering could be made available.

A new family of broadband wireless technologies has emerged that poses a threat to traditional wired access infrastructures. These include *Local Multipoint Distribution Service* (LMDS) and *Multichannel Multipoint Distribution Service* (MMDS), as well as *Geosynchronous* (GEO) and *Low Earth Orbit* (LEO) satellites.

Local Multipoint Distribution Service (LMDS)

LMDS is a bottleneck resolution technology designed to alleviate the transmission restriction that occurs between high-speed LANs and WANs. Today, local networks routinely operate at speeds of 100 Mbps (Fast Ethernet) and even 1,000 Mbps (Gigabit Ethernet), which means that any local loop solution that operates slower than either of those poses a restrictive barrier to the overall performance of the system. LMDS offers a good alternative to wired options. Originally offered as CellularVision, it was seen by its inventor, Bernard Bossard, as a way to provide cellular television as an alternative to cable.

Operating in the 28 GHz range, LMDS offers data rates as high as 155 Mbps, the equivalent of SONET OC-3c. Because it is a wireless solution, it requires a minimal infrastructure and can be deployed quickly and cost effectively as an alternative to the wired infrastructure provided by incumbent service providers. After all, the highest-cost component (more than 50 percent) when building networks is not the distribution facility, but rather the labor required to trench it into the ground or build aerial facilities. Thus, any access alternative that minimizes the cost of labor will garner significant attention.

LMDS relies on a cellular-like deployment strategy under which the cells are approximately three miles in diameter. Unlike cellular service, however, users are stationary. Consequently, LMDS cells do not need to support roaming. Antenna/transceiver units are generally placed on rooftops, as they need an unobstructed line of sight to operate properly. In fact, this is one of the disadvantages of LMDS (and a number of other wireless technologies): Besides suffering from unexpected physical obstructions, the service suffers from "rain fade" caused by absorption and the scattering of the transmitted microwave signal by atmospheric moisture. Even some forms of foliage will cause interference for LMDS, so the transmission and reception equipment must be mounted high enough to avoid such obstacles, hence the tendency to mount the equipment on rooftops.

Because of its high bandwidth capability, many LMDS implementations interface directly with an ATM backbone to take advantage of both its bandwidth and its diverse QoS capability. If ATM is indeed the transport fabric of choice, then the LMDS service becomes a broadband access alternative to a network capable of transporting a full range of services including voice, video, image, and data—the full suite of multimedia applications.

Multichannel, Multipoint Distribution System (MMDS)

MMDS got its start as a wireless cable television solution. In 1963, a spectrum allocation known as the *Instructional Television Fixed Service* (ITFS) was carried out by the FCC as a way to distribute educational content to schools and universities. In the 1970s, the FCC established a two-channel metropolitan distribution service called the *Multipoint Distribution Service* (MDS). It was to be used for the delivery of pay TV signals, but with the advent of inexpensive satellite access and the ubiquitous deployment of cable systems, the need for MDS went away.

In 1983, the FCC rearranged the MDS and ITFS spectrum allocation, creating 20 ITFS education channels and 13 MDS channels. In order to qualify to use the ITFS channels, schools had to use a minimum of 20 hours of airtime, which meant that ITFS channels tended to be heavily, albeit randomly, utilized. As a result, MMDS providers that use all 33 MDS and ITFS channels must be able to dynamically map requests for service to available channels in a completely transparent fashion, which means that the bandwidth management system must be reasonably sophisticated.

Because MMDS is not a true cable system (in spite of the fact that it has its roots in television distribution), no franchise issues occupy its use (of course, certain licensing requirements must be met). However, the technology is also limited in terms of what it can do. Unlike LMDS, MMDS is designed as a one-way broadcast technology and therefore does not typically enable upstream communication. Many contend, however, that MMDS contains the adequate bandwidth to provision two-way systems, which would make it suitable for voice, Internet access, and other data-oriented services.

So What's the Market?

The largest opportunity for fixed wireless technology is in the metro environment due to such an area being able to overcome congested conduit, the permits required to dig up streets to lay more infrastructure, and technology cost. Service providers looking to sell LMDS and MMDS technologies should target small and medium-size businesses that experience measurable peak data rates and that are looking to move into the packet-based transport arena, many of which are found in metro areas. Typical applications for the technologies include LAN interconnections, Internet access, and cellular backhaul between *mobile telephone switching offices* (MTSOs) and newly deployed cell sites. LMDS tends to offer higher data rates than MMDS. LMDS peaks out at a whopping 1.5 Gbps, while MMDS can achieve a maximum transmission speed of about 3 Mbps. Nevertheless, both have their place in the technology pantheon.

Summary

Access technologies, used to connect the customer to the network, come in a variety of forms and offer a broad variety of connectivity options and bandwidth levels. The key to success is to *not* be a bottleneck; access technologies that can evolve to meet the growing customer demands for bandwidth will be the winners in the game. DSL holds an advantage as long as it can overcome the availability challenge and the technology challenge of loop-carrier restrictions. Wireless is hobbled by licensing and spectrum availability, both of which are regulatory and legal in nature rather than being technology limitations.

Transport Technologies

We have now discussed the premises environment and access technologies. The next area we'll examine is *transport*.

Because businesses are rarely housed in a single building, and because their customers are often scattered across a broad metropolitan area, a growing need exists for high-speed, reliable wide area transport. *Wide area* can take on a variety of meanings. For example, a company with multiple offices scattered across the metropolitan expanse of a large city requires interoffice connectivity in order to do business properly. On the other hand, a large multinational with offices and clients in four global cities requires connectivity to ensure that the offices can exchange information on a 24-hour basis. In either case, the service must be the same. The users on either end of the connection must be locally connected, whether they are on opposite sides of a building or on opposite sides of the planet.

Although global long-haul connectivity really isn't a metro technology, it *is* a metro application, because a metropolitan area-based company may very well have far-flung offices that it must connect to. Furthermore, the word metro has become a bit fuzzy. Certain metro networks span metropolitan Los Angeles, a distance of well over 100 miles.

These requirements are satisfied through the proper deployment of wide area transport technologies. These can be as simple as a dedicated private-line circuit or as complex as a virtual installation that relies on ATM for high-quality transport.

Dedicated facilities are excellent solutions because they are dedicated. They provide fixed bandwidth that never varies and guarantee the quality of the transmission service. Because they are dedicated, however, they suffer from two disadvantages. First, they are expensive and only cost-effective when highly utilized. The pricing model for dedicated circuits includes two components: the mileage of the circuit and the bandwidth. The longer the circuit, and the faster it is, the more it costs. Second, because they are not switched and are often not redundant because of cost, dedicated facilities pose the potential threat of a prolonged service outage should they fail.

Nevertheless, dedicated circuits are popular for certain applications and are widely deployed. They include such solutions as T1, which offers 1.544 Mbps of bandwidth; DS-3, which offers 44.736 Mbps of bandwidth; and SONET, which offers a wide range of bandwidth from 51.84 Mbps to as much as 40 Gbps.

The alternative to a dedicated facility is a switched service, such as Frame Relay or ATM. These technologies provide virtual circuits. Instead of dedicating physical facilities, they dedicate logical timeslots to each customer who then shares access to physical network resources. In the case of Frame Relay, the service can provide bandwidth as high as DS-3, thus providing an ideal replacement technology for lower-speed dedicated circuits. ATM, on the other hand, operates hand in glove with SONET and is thus capable of providing transport services at gigabit speeds. Finally, the new field of optical networking is carving out a large niche for itself as a bandwidth-rich solution with the potential for inherent QoS.

We begin our discussion with dedicated private line, otherwise known as point-to-point.

Point-to-Point Technologies

Point-to-point technologies do exactly what their name implies. They connect one point directly with another, as shown in Figure 3-20. For example, it is common for two buildings in a downtown area to be connected by a point-to-point microwave or infrared circuit because the cost of establishing it is far lower than the cost to put in physical facilities in a crowded city. Many businesses rely on dedicated, point-to-point optical facilities to interconnect locations, especially businesses that require dedicated bandwidth for high-speed applications. Of course, point-to-point does not necessarily imply high bandwidth; many locations use 1.544 Mbps T1 facilities for interconnections, and some rely on lower-speed circuits where higher bandwidth is not required.

Figure 3-20
A point-to-point circuit inter-connecting two distinct endpoints

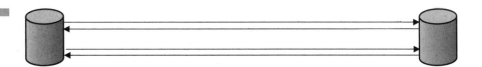

Dedicated facilities provide bandwidth from as low as 2,400 bits per second (rare) to as high as multiple gigabits per second. The low-rate analog facilities are often used for alarm circuits and telemetry, while circuits operating at 4,800 and 9,600 *bits per second* (bps) are used to access interactive, host-based data applications.

Higher-speed facilities are usually digital and are often channelized by dedicated multiplexers and shared among a collection of users or by a variety of applications. For example, a high-bandwidth facility that interconnects two corporate locations might be dynamically subdivided into various sized channels to be used by a PBX for voice, a videoconferencing system, and data traffic.

Dedicated facilities have the advantage of always being available to the subscriber, but they have the disadvantage of accumulating charges regardless of whether they are actually in use. For the longest time, dedicated circuits represented the only solution that provided guaranteed bandwidth; switched solutions simply weren't designed for the heavy service requirements of graphical- and data-intensive traffic. Over time, however, that has changed. A number of switched solutions have emerged in the last few years that provide guaranteed bandwidth and only accumulate charges when they are being used (although some of them offer very reasonable fixed-rate service). The two most common of these are Frame Relay and ATM; we begin with Frame Relay.

Frame Relay   Frame Relay came about as a private-line replacement technology and was originally intended as a data-only service. Today, it carries not only data, but voice and video as well, and although it was originally crafted with a top speed of T1/E1, it now provides connectivity at much higher bandwidth levels.

In Frame Relay networks, the incoming data stream is packaged as a series of variable-length frames that can transport any kind of data: LAN traffic, IP packets, SNA frames, and even voice and video. In fact, it has been recognized as a highly effective transport mechanism for voice, allowing Frame Relay-capable PBXs to be connected to a Frame Relay *permanent virtual circuit* (PVC), which can cost-effectively replace private-line circuits used for the same purpose. When voice is carried over Frame Relay, it is usually compressed for transport efficiency and packaged in small frames to minimize the processing delay of the frames. According to the Frame Relay Forum, as many as 255 voice channels can be encoded over a single PVC, although the number is usually smaller when actually implemented.

Frame Relay is a virtual circuit service. When customers want to connect two locations using Frame Relay, they contact their service provider and tell the service representative the locations of the endpoints and the bandwidth they require. The service provider issues a service order to create the circuit. If, at some point in the future, the customer decides to change the circuit endpoints or upgrade the bandwidth, a PVC service order must be issued, which is the most commonly deployed Frame Relay solution.

Frame Relay is also capable of supporting *switched virtual circuit* (SVC) service, but SVCs are for the most part not available from service providers. With SVC service, customers can make their own modifications to the circuit by accessing the Frame Relay switch in the central office and requesting changes. However, service providers do not currently offer SVC service because of billing and tracking concerns (customer activities are difficult to monitor). Instead, they allow customers to create a fully meshed network between all locations for a very reasonable price. Instead of making routing changes in the switch, the customer has a circuit between every possible combination of desired endpoints. As a result, customers get the functionality of a switched network, while the service provider avoids the difficulty of administering a network within which the customer is actively making changes.

In Frame Relay, PVCs are identified using an address called a *Data Link Connection Identifier* (DLCI, pronounced *delsie*). At any given endpoint, the customer's router can support multiple DLCIs, and each DLCI can be assigned varying bandwidths based upon the requirements of the device/application on the router port associated with that DLCI.

In Figure 3-21, the customer has purchased a T1 circuit to connect their router to the Frame Relay network. The router is connected to a videoconferencing unit at 384 Kbps, a Frame Relay-capable PBX at 768 Kbps, and a data circuit for Internet access at 512 Kbps. Note that the aggregate bandwidth assigned to these devices exceeds the actual bandwidth of the access line by 128 Kbps (1,664 to 1,536). Under normal circumstances this would not be possible, but Frame Relay assumes that the traffic that it will normally be transporting is bursty by nature.

If the assumption is correct (and it usually is), it is unlikely that all three devices will burst at the same instant in time. As a consequence, the circuit's operating capacity can actually be overbooked, a process known as *oversubscription.* Most service providers allow as much as 200 percent oversubscription, something customers clearly benefit from, provided the circuit is designed properly. This means that the salesperson must carefully assess the nature of the traffic that the customer will be sending over the link and ensure that enough bandwidth is allocated to support the requirements of the various devices that will be sharing access to the link. Failure to do so can result in an underengineered facility that will not meet the customer's

Figure 3-21
In this Frame Relay
service delivery, the
customer has
provisioned a T1
circuit from his or her
premises network to
the Frame Relay
cloud.

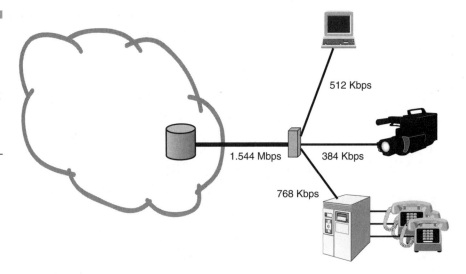

512 Kbps

1.544 Mbps   384 Kbps

768 Kbps

throughput requirements. This is a critical component of the service delivery formula.

The throughput level, that is, the bandwidth that Frame Relay service providers absolutely guarantee on a PVC-by-PVC basis, is called the *Committed Information Rate* (CIR). In addition to CIR, service providers will often support an *Excess Information Rate* (EIR), which is the rate above the CIR they will attempt to carry, assuming the capacity is available within the network. However, all frames above the CIR are marked as eligible for discard, which simply means that the network will do its best to deliver them, but it makes no guarantees. If push comes to shove, and the network finds itself to be congested, the frames marked *discard eligible* (DE) are immediately discarded at their point of ingress.

This CIR/EIR relationship is poorly understood by many customers because the CIR is taken to be an indicator of the absolute bandwidth of the circuit. Whereas bandwidth is typically measured in bps, CIR is a measure of *bits in one second.* In other words, the CIR is a measure of the average throughput that the network will guarantee. The actual transmission volume of a given CIR may be higher or lower than the CIR at any point in time because of the bursty nature of the data being sent, but in aggregate the network will maintain an average, guaranteed flow volume for each PVC. This is a selling point for Frame Relay.

In most cases, customers get more than they actually pay for, and as long as the switch loading levels are properly engineered, the switch (and therefore the Frame Relay service offering) will not suffer adversely from this charitable bandwidth allocation philosophy. The key to success when selling Frame Relay is to have a very clear understanding of the applications the customer intends to use across the link so that the access facility can be properly sized for the anticipated traffic load.

Managing Service in Frame Relay Networks   Frame Relay does not offer a great deal of granularity when it comes to QoS. The only inherent mechanism is the DE bit described earlier as a way to control network congestion. However, the DE bit is binary. It has two possible values, which means that a customer has two choices: The information being sent is either important or it isn't, and this is not particularly useful for establishing a variety of QoS levels.

Consequently, a number of vendors have implemented proprietary solutions for QoS management. Within their routers (sometimes called *Frame Relay Access Devices* [FRADs]) they have established queuing mechanisms that enable customers to create multiple priority levels for differing traffic flows. For example, voice and video, which don't tolerate delay well, could be assigned to a higher-priority queue than the one to which asynchronous data traffic would be assigned. This allows Frame Relay to provide highly granular service. The downside is that this approach is proprietary, which means that the same vendor's equipment must be used on both ends of the circuit. Given the strong move toward interoperability, this is not an ideal solution because it locks the customer into a single-vendor situation.

Congestion Control in Frame Relay   Frame Relay has two congestion control mechanisms. Embedded in the header of each Frame Relay frame are two additional bits called the *Forward Explicit Congestion Notification* (FECN) bit and the *Backward Explicit Congestion Notification* (BECN) bit. Both are used to notify devices in the network of congestion that could affect throughput.

Consider the following scenario. A Frame Relay frame arrives at the second of three switches along the path to its intended destination, where it encounters severe local congestion (see Figure 3-22). The congested switch sets the FECN bit to indicate the presence of congestion and transmits the frame to the next switch in the chain. When the frame arrives, the receiving switch takes note of the FECN bit, which tells it the following: "I just came from that switch back there, and it's extremely congested. You can transmit stuff back there if you want to, but there's a good chance that anything you send will be discarded, so you might want to wait awhile before transmitting." In other words, the switch has been notified of a congestion

Figure 3-22

A frame encounters congestion in transit. Because of local congestion, the network sets the FECN bit on the outbound frame.

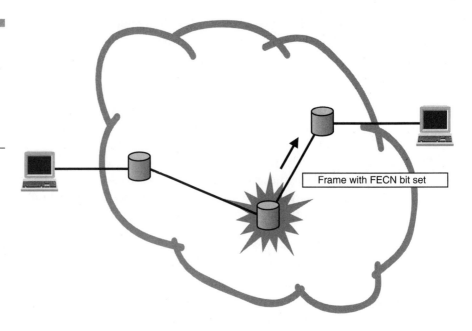

Frame with FECN bit set

condition, to which it may respond by throttling back its output to enable the affected switch time to recover.

On the other hand, the BECN bit is used to flow control a device that is sending too much information into the network. Consider the situation shown in Figure 3-23 where a particular device on the network is transmitting at a high volume level, routinely violating the CIR, and perhaps the EIR level established by mutual consent. The ingress switch—that is, the first switch the traffic touches—has the capability to set the BECN bit on frames going toward the offending device, and the bit carries the implicit message, "Cut it out or I'm going to hurt you." In effect, the BECN bit notifies the offending device that it is violating protocol, and continuing to do so will result in every frame from that device being discarded without warning or notification. If this happens, it gives the ingress switch the opportunity to recover. However, it doesn't fix the problem; it merely forestalls the inevitable, because sooner or later the intended recipient will realize that frames are missing and will initiate recovery procedures, which will cause resends to occur. However, it may give the affected devices time to recover before the onslaught begins anew.

The problem with FECN and BECN lies in the fact that many devices choose not to implement them. They do not necessarily have the inherent capability to throttle back upon receipt of a congestion indicator, although

Figure 3-23
A BECN bit used to
warn a station that it
is transmitting too
much data (in
violation of its
agreement with
the switch)

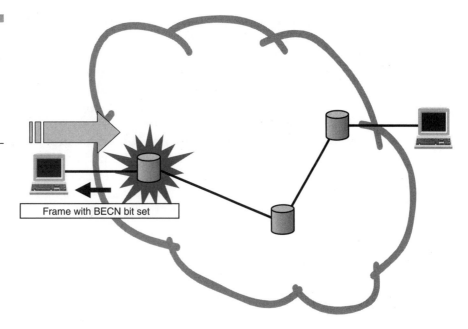

Frame with BECN bit set

devices that can are becoming more common. Nevertheless, proprietary solutions are in widespread use and will continue to be so for some time to come.

Frame Relay Summary   Frame Relay evolved quickly from a data-only transport scheme to a multiservice technology with diverse capabilities. For data and some voice and video applications, it shines as a WAN offering. In some areas, however, Frame Relay is lacking. Its bandwidth is limited to DS-3, and its capability to offer standards-based QoS is limited. Given the focus on QoS that is so much a part of customers' chanted mantra today, and the flexibility that a switched solution permits, something else is required. That something is ATM.

Asynchronous Transfer Mode (ATM)

Network architectures often develop in concert with the corporate structures that they serve. Companies with centralized management authorities such as utilities, banks, and hospitals often have centralized and tightly controlled hierarchical data processing architectures to protect their data. On the other hand, organizations that are distributed in nature such as research and development facilities and universities often have highly distributed data-processing architectures. They tend to share information on a peer-to-peer basis and their corporate structures reflect this fact.

ATM came about not only because of the proliferation of diverse network architectures, but also because of the evolution of traffic characteristics and transport requirements. To the well-known demands of voice, we now add various flavors of data, video, MP3, an exponentially large variety of IP traffic, interactive real-time gaming, and a variety of other content types that place increasing demands on the network. Further, we have seen a requirement arise for a mechanism that can transparently and correctly transport the mix of various traffic types over a single network infrastructure while at the same time delivering granular, controllable, and measurable QoS levels for each service type. In its original form, ATM was designed to do exactly that, work with SONET or SDH to deliver high-speed transport and switching throughout the network: in the wide area, the metropolitan area, the campus environment, and the LAN, right down to the desktop, seamlessly, accurately, and fast.

Today, because of competition from such technologies as QoS-aware IP transport, proprietary high-speed mesh networks, and Fast and Gigabit Ethernet, ATM has for the most part lost the race to the desktop. ATM is a cell-based technology, which simply means that the fundamental unit of transport—a frame of data, if you will—is of a fixed size. This enables switch designers to build faster, simpler devices, because they can always count on their switched payload being the same size at all times. That cell comprises a five-octet header and a 48-octet payload field, as shown in Figure 3-24.

The payload contains user data and the header contains information that the network requires to both transport the payload correctly and ensure proper QoS levels for the payload. ATM accomplishes this task well, but at a cost. The five-octet header comprises nearly 10 percent of the cell,

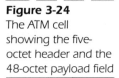

Figure 3-24
The ATM cell showing the five-octet header and the 48-octet payload field

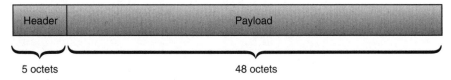

Header	Payload

5 octets   48 octets

a rather significant price to pay, particularly when other technologies such as IP and SONET add their own significant percentages of overhead to the overall payload. This reality is part of the problem. ATM's original claims to fame, and the reasons it rocketed to the top of the technology hit parade, were its capability to switch cells at a tremendous speed through the WAN and the ease with which the technology could be scaled to fit any network situation. Today, however, given the availability of high-speed IP routers that routinely route packets at terabit rates, ATM's advantages have begun to pale to a certain degree.

ATM Evolution

ATM has, however, emerged from the flames in other ways. Today, many service providers see ATM as an ideal aggregation technology for diverse traffic streams that need to be combined for transport across a WAN that will most likely be IP-based. ATM devices, then, will be placed at the edge of the network, where they will collect traffic for transport across the Internet or (more likely) a privately owned IP network. Furthermore, because it has the capability to be something of a chameleon by delivering diverse services across a common network fabric, it is further guaranteed a seat at the technology game.

It is interesting to note that the traditional, legacy telecommunications network comprises two principal regions that can be clearly distinguished: the network itself, which provides switching, signaling, and transport for traffic generated by customer applications; and the access loop, which provides the connectivity between the customer's applications and the network. In this model, the network is considered to be a relatively intelligent medium, while the customer equipment is usually considered to be relatively "stupid."

Not only is the intelligence seen as being concentrated within the confines of the network, so too is the bulk of the bandwidth since the legacy model indicates that traditional customer applications don't require much of it. Between central office switches, however, and between the offices themselves, enormous bandwidth is required.

Today, this model is changing. Customer equipment has become remarkably intelligent, and many of the functions previously done within the network cloud are now performed at the edge. PBXs, computers, and other devices are now capable of making discriminatory decisions about required service levels, eliminating any need for the massive intelligence embedded in the core.

At the same time, the bandwidth is migrating from the core of the network toward the customer as applications evolve to require it. Massive bandwidth still exists within the cloud, but the margins of the cloud are expanding toward the customer.

The result of this evolution is a redefinition of the network's regions. Instead of a low-speed, low-intelligence access area and a high-speed, highly intelligent core, the intelligence has migrated outward to the margins of the network and the bandwidth, once exclusively a core resource, is now equally distributed at the edge. Thus, we see something of a core and edge distinction evolving as customer requirements change. This clearly has *major* implications for the metro customer and service provider alike, because the metro environment lies at the margins of the core network, often serving as the interface point between access and high-speed transport.

One reason for this steady migration is the well-known fact within sales and marketing circles that products sell best when they are located close to the buying customer. They are also easier to customize for individual customers when they are physically closest to the situation for which the customer is buying them.

ATM Technology Overview

Because ATM plays such a major role in networks today it is important to develop at least a rudimentary understanding of its functions, architectures, and offered services.

ATM Protocols

Like all modern technologies, ATM has a well-developed protocol stack, shown in Figure 3-25, which clearly delineates the functional breakdown of the service. The stack consists of four layers: the upper services layer, the *ATM adaptation layer* (AAL), the ATM layer, and the physical layer.

The upper services layer defines the nature of the actual services that ATM can provide. It identifies both constant and variable bit rate services. Voice is an example of a constant bit rate service, while signaling, IP, and Frame Relay are examples of both connectionless and connection-oriented variable bit rate services.

The AAL has four general responsibilities:

- Synchronization and recovery from errors
- Error detection and correction

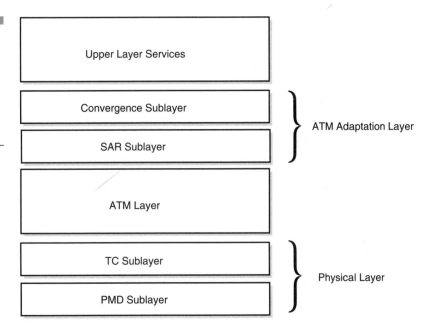

Figure 3-25
The ATM protocol stack. As with all layered protocols, each layer is dependent on the layers surrounding it for functionality.

◼ Segmentation and reassembly of the data stream

◼ Multiplexing

The AAL comprises two functional sublayers. The *convergence sublayer* provides service-specific functions to the services layer, which can then make the most efficient use of the underlying cell relay technology that ATM provides. Its functions include clock recovery for end-to-end timing management, a recovery mechanism for lost or out-of-order cells, and a timestamp capability for time-sensitive traffic such as voice and video.

The *segmentation and reassembly sublayer* (SAR) converts the user's data from its original incoming form into the 48-octet payload "chunks" that will become cells. For example, if the user's data is arriving in the form of IP packets, SAR chops them into 48-octet payload pieces. It also has the responsibility to detect lost or out-of-order cells that the convergence sublayer will recover from, and it can detect single bit errors in the payload chunks.

The ATM layer has five general responsibilities:

◼ Cell multiplexing and demultiplexing

◼ Virtual path and virtual channel switching

- Creation of the cell header
- Generic flow control
- Cell delineation

Because the ATM layer creates the cell header, it is responsible for all the functions that the header manages. The process, then, is fairly straightforward: The user's data passes from the services layer to the ATM adaptation layer, which segments the data stream into 48-octet pieces. The pieces are handed to the ATM layer, which creates the header and attaches it to the payload unit, thus creating a cell. The cells are then handed down to the physical layer.

The physical layer consists of two functional sublayers as well: the transmission convergence sublayer and the physical medium sublayer. The transmission convergence sublayer performs three primary functions. The first is called cell rate decoupling, which adapts the cell creation and transmission rate to the rate of the transmission facility by performing *cell stuffing*, similar to the bit-stuffing process described earlier in the discussion of DS-3 frame creation. The second responsibility is cell delineation, which enables the receiver to delineate between one cell and the next. Finally, it generates the transmission frame in which the cells are to be carried.

The physical medium sublayer takes care of issues that are specific to the medium being used for transmission, such as line codes, electrical and optical concerns, timing, and signaling.

The physical layer can use a wide variety of transport options, including

- DS1/DS2/DS3
- E1/E3
- 25.6 Mbps *User-to-Network Interface* (UNI) over UTP-3
- 51 Mbps UNI over UTP-5 (*Transparent Asynchronous Transmitter/Receiver Interface* [TAXI])
- 100 Mbps UNI over UTP-5
- OC3/12/48c

Others, of course, will follow as transport technologies advance.

The ATM Cell Header

As we mentioned before, ATM is a cell-based technology that relies on a 48-octet payload field that contains actual user data and a 5-byte header that

Figure 3-26
ATM cell header
details

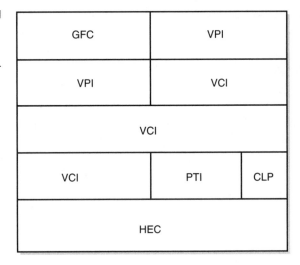

GFC	VPI
VPI	VCI
VCI	
VCI	PTI	CLP
HEC	

contains information needed by the network to route the cell and provide proper levels of service.

The ATM cell header, shown in Figure 3-26, is examined and updated by each switch it passes through. The header comprises six distinct fields: the *Generic Flow Control* (GFC) field, the *Virtual Path Identifier* (VPI) field, the *Virtual Channel Identifier* (VCI) field, the *Payload Type Identifier* (PTI) field, the *Cell Loss Priority* (CLP) field, and the *Header Error Control* (HEC) field:

- **GFC field** This four-bit field is used across the UNI for network-to-user flow control. It has not yet been completely defined in the ATM standards, but some companies have chosen to use it for very specific purposes.

- **VPI field** The eight-bit VPI identifies the virtual path over which the cells will be routed at the UNI. It should be noted that because of dedicated, internal flow control capabilities within the network, the GFC field is not needed across the *network-to-network interface* (NNI). It is therefore redeployed; the four bits are converted to additional VPI bits, thus extending the size of the virtual path field. This allows for the identification of more than 4,000 unique VPs. At the UNI, this number is excessive, but across the NNI it is necessary because of the number of potential paths that might exist between the switches that make up the fabric of the network.

■ **VCI field**   As the name implies, the 16-bit VCI identifies the unidirectional virtual channel over which the current cells will be routed.

■ **PTI field**   The three-bit PTI field is used to indicate network congestion and cell type, in addition to a number of other functions. The first bit indicates whether the cell was generated by the user or by the network, while the second indicates the presence or absence of congestion in user-generated cells or flow-related *Operations, Administration, and Maintenance* (OAM) information in cells generated by the network. The third bit is used for service-specific, higher-layer functions in the user-to-network direction, such as indicating that a cell is the last in a series of cells. From the network to the user, the third bit is used with the second bit to indicate whether the OAM information refers to segment or end-to-end-related information flow.

■ **CLP field**   This single-bit field is a relatively primitive flow control mechanism by which the user can indicate to the network which cells to discard in the event of a condition that demands some cells be eliminated, similar to the DE bit in Frame Relay. It can also be set by the network to indicate to downstream switches that certain cells in the stream are eligible for discard, should that become necessary.

■ **HEC field**   The eight-bit HEC field can be used for two purposes. First, it provides for the calculation of an eight-bit CRC that checks the integrity of the entire header. Second, it can be used for cell delineation.

Addressing in ATM

ATM is a connection-oriented, virtual circuit technology, meaning that communication paths are created through the network prior to actually sending traffic. Once established, the ATM cells are routed based upon a virtual circuit address. A virtual circuit is simply a connection that gives the user the appearance of being dedicated to that user, when in point of fact the only thing that is actually dedicated is a time slot. This technique is generically known as *label-based switching* and is accomplished through the use of routing tables in the ATM switches that designate input ports, output ports, input addresses, output addresses, and QoS parameters required for proper routing and service provisioning. As a result, cells do not contain explicit destination addresses but rather contain timeslot identifiers.

Figure 3-27
Addressing in ATM with virtual channels and paths. Virtual channels are unidirectional; a virtual path is made up of two or more virtual channels.

Every virtual circuit address has two components, as shown in Figure 3-27. The first is the virtual channel, which is a unidirectional conduit for the transmission of cells between two endpoints. For example, if two parties are conducting a videoconference, they will each have a virtual channel for the transmission of outgoing cells that make up the video portion of the conference.

The second level of the ATM addressing scheme is called a virtual path. A virtual path is a bundle of virtual channels that have the same endpoints and that, when considered together, make up a bidirectional transport facility. The combination of unidirectional channels that we need in our two-way videoconferencing example makes up a virtual path.

ATM Services

The basic services that ATM provides are based on three general characteristics: the nature of the connection between the communicating stations (connection-oriented versus connectionless), the timing relationship between the sender and the receiver, and the bit rate required to ensure proper levels of service quality. Based on those generic requirements, both the *International Telecommunication Union—Telecommunications Standardization Sector* (ITU-T) and the ATM Forum have created service classes that address the varying requirements of the most common forms of transmitted data.

ITU-T Service Classes The ITU-T assigns service classes based on three characteristics: the connection mode, bit rate, and the end-to-end timing relationship between the end stations. They have created four distinct service classes, based on the model shown in Figure 3-28. Class A service, for

Figure 3-28
ITU-T ATM service
definitions

	Class A	Class B	Class C	Class D
AAL Type	1	2	5, 3/4	5, 3/4
Connection Mode	Connection-oriented	Connection-oriented	Connection-oriented	Connectionless
Bit Rate	Constant	Variable	Variable	Variable
Timing relationship	Required	Required	Not required	Not required
Service Types	Voice, video	VBR voice, video	Frame relay	IP

example, defines a connection-oriented, constant-bit-rate, timing-based service that is ideal for the stringent requirements of voice service. Class B, on the other hand, is ideal for such services as variable bit rate video in that it defines a connection-oriented, variable-bit-rate, timing-based service.

Class C service was defined for such things as Frame Relay in that it provides a connection-oriented, variable-bit-rate, timing-independent service. Finally, Class D delivers a connectionless, variable-bit-rate, timing-independent service that is ideal for IP traffic as well as *Switched Multimegabit Data Service* (SMDS).

In addition to service classes, the ITU-T has defined AAL service types, which align closely with the A, B, C, and D service types described previously. Whereas the service classes (A, B, C, and D) describe the capabilities of the underlying network, the AAL types describe the cell format. They are AAL1, AAL2, AAL3/4, and AAL5. However, only two of them have really survived in a big way.

AAL1 is defined for Class A service, which is a constant bit rate environment ideally suited for voice and voice-like applications. In AAL1 cells, the first octet of the payload serves as a payload header that contains cell sequence and synchronization information that is required to provision the constant bit rate, fully sequenced service. AAL1 provides circuit emulation service without dedicating a physical circuit, which explains the need for an end-to-end timing relationship between the transmitter and the receiver.

AAL5, on the other hand, is designed to provide both Class C and D services, and although it was originally proposed as a transport scheme for connection-oriented data services, it turns out to be more efficient than AAL3/4 and accommodates connectionless services quite well.

To guard against the possibility of errors, AAL5 has an eight-octet trailer appended to the user data. The trailer includes a variable size Pad field used to align the payload on 48-octet boundaries, a two-octet Control field that is currently unused, a two-octet Length field that indicates the number of octets in the user data, and finally a four-octet CRC that can check the integrity of the entire payload. AAL5 is often referred to as the *simple and easy adaptation layer* (SEAL), and it may find an ideal application for itself in the burgeoning Internet arena. Recent studies indicate that *Transmission Control Protocol* (TCP)/IP transmissions produce comparatively large numbers of small packets that tend to be around 48 octets long. That being the case, AAL5 could well transport the bulk of them in its user data field. Furthermore, the maximum size of the user data field is 65,536 octets, coincidentally the same size as an IP packet.

ATM Forum Service Classes The ATM Forum looks at service definitions slightly differently than the ITU-T, as shown in Figure 3-29. Instead of the A-B-C-D services, the ATM Forum categorizes them as real-time and nonreal-time services. Under the real-time category, they define constant bit rate services that demand fixed resources with guaranteed availability. They also define real-time variable bit rate service, which provides for statistical multiplexed, variable bandwidth service allocated on demand. A further subset of real-time VBR is peak-allocated VBR, which guarantees constant loss and delay characteristics for all cells in that flow.

Under the nonreal-time service class, *Unspecified Bit Rate* (UBR) is the first service category. UBR is often compared to IP in that it is a "best-effort"

Figure 3-29

ATM Forum service definitions

Service	Descriptors	Loss	Delay	Bandwidth	Feedback
CBR	PCR, CDVT	Yes	Yes	Yes	No
VBR-RT	PCR, CDVT, SCR, MBS	Yes	Yes	Yes	No
VBR-NRT	PCR, CDVT, SCR, MBS	Yes	Yes	Yes	No
UBR	PCR, CDVT	No	No	No	No
ABR	PCR, CDVT, MCR	Yes	No	Yes	Yes

delivery scheme in which the network provides whatever bandwidth it has available, with no guarantees made. All recovery functions from lost cells are the responsibility of the end-user devices.

UBR has two subcategories of its own. The first, *nonreal-time VBR* (NRT-VBR), improves the impacts of cell loss and delay by adding a network resource reservation capability. *Available bit rate* (ABR), UBR's other subcategory, makes use of feedback information from the far end to manage loss and ensure fair access to and transport across the network.

Each of the five classes makes certain guarantees with regard to cell loss, cell delay, and available bandwidth. Furthermore, each of them takes into account descriptors that are characteristic of each service described. These include *peak cell rate* (PCR), *sustained cell rate* (SCR), *minimum cell rate* (MCR), *cell delay variation tolerance* (CDVT), and *burst tolerance* (BT).

ATM-Forum-Specified Services   The ATM Forum has identified a collection of services for which ATM is a suitable, perhaps even desirable, network technology. These include *cell relay service* (CRS), *circuit emulation service* (CES), *voice and telephony over ATM* (VTOA), *Frame Relay bearer service* (FRBS), *LAN emulation* (LANE), *multiprotocol over ATM* (MPOA), and a collection of others.

CRS is the most basic of the ATM services. It delivers precisely what its name implies: a "raw pipe" transport mechanism for cell-based data. As such, it does not provide any ATM bells and whistles, such as QoS discrimination; nevertheless, it is the most commonly implemented ATM offering because of its lack of implementation complexity.

CES gives service providers the ability to offer a selection of bandwidth levels by varying both the number of cells transmitted per second and the number of bytes contained in each cell.

VTOA is a service that has yet to be clearly defined. The capability to transport voice calls across an ATM network is a nonissue, given the availability of Class A service. What is not clearly defined, however, are corollary services such as 800/888 calls, 900 service, 911 call handling, enhanced services billing, *Signaling System 7* (SS7) signal interconnection, and so on. Until these issues are clearly resolved, ATM-based, feature-rich telephony will not become a mainstream service but will instead be limited to simple voice—and there *is* a difference.

FRBS refers to ATM's capability to interwork with Frame Relay. Conceptually, the service implies that an interface standard allows an ATM switch to exchange data with a Frame Relay switch, thus allowing for interoperability between frame and cell-based services. Many manufacturers are taking a slightly different tack, however: They build switches with soft, chewy cell technology at the core and surround the core with hard, crunchy interface cards to suit the needs of the customer.

For example, an ATM switch might have ATM cards on one side to interface with other ATM devices on the network and Frame Relay cards on the other side to allow it to communicate with other Frame Relay switches, as shown in Figure 3-30. Thus, a single piece of hardware can logically serve as both a cell and Frame Relay switch. This design is becoming more and more common, because it helps to avoid a future rich with forklift upgrades.

LANE enables an ATM network to move traffic transparently between two similar LANs but also to allow ATM to transparently slip into the LAN arena. For example, two Ethernet LANs could communicate across the fabric of an ATM network, as could two token ring LANs. In effect, LANE enables ATM to provide a bridging function between similar LAN environments. In LANE implementations, the ATM network does not handle MAC functions such as collision detection, token passing, or beaconing; it merely provides the connectivity between the two communicating endpoints. The MAC frames are simply transported inside AAL5 cells.

One clear concern about LANE is that LANs are connectionless, while ATM is a virtual-circuit-based, connection-oriented technology. LANs routinely broadcast messages to all stations, while ATM enables point-to-point or multipoint circuits only. Thus, ATM must look like a LAN if it is to behave like one. To make this happen, LANE uses a collection of specialized

Figure 3-30
In FRBS, ATM emulates Frame Relay service. This is a good example of technology and services convergence.

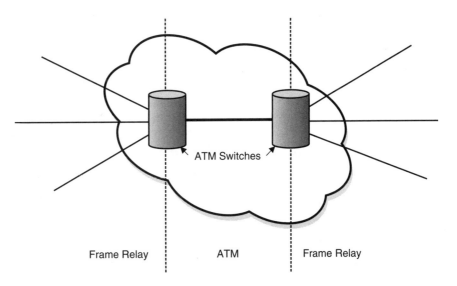

ATM Switches

Frame Relay · ATM · Frame Relay

LANE clients and servers to provide the connectionless behavior expected from the ATM network.

On the other hand, MPOA provides the ATM equivalent of *routing* in LAN environments. In MPOA installations, routers are referred to as MPOA servers. When one station wants to transmit to another station, it queries its local MPOA server for the remote station's ATM address. The local server then queries its neighbor devices for information about the remote station's location. When a server finally responds, the originating station uses the information to establish a connection with the remote station, while the other servers cache the information for further use.

MPOA promises a great deal, but it is complex to implement and requires other ATM components, such as the Private NNI capability, to work properly. Furthermore, it's being challenged by at least one alternative technology, known as IP switching.

IP switching is far less overhead intensive than MPOA. Furthermore, it takes advantage of a known (but often ignored) reality in the LAN interconnection world: Most routers today use IP as their core protocol, and the great majority of LANs are still Ethernet. This means that a great deal of simplification can be done by crafting networks to operate around these two technological bases, and in fact, this is precisely what IP switching does. By using existing, low-overhead protocols, the IP switching software creates new ATM connections dynamically and quickly, updating switch tables on the fly. In IP switching environments, IP resides on top of ATM within a device, as shown in Figure 3-31, providing the best of both protocols. If two communicating devices want to exchange information and they have done so before, an ATM mapping already exists and no layer three involvement (IP) is required; the ATM switch portion of the service simply creates the connection at high speed. If an address lookup is required, then the "call" is handed up to IP, which takes whatever steps are required to perform the lookup (a DNS request, for example). Once it has the information, it hands it down to ATM, which proceeds to set up the call. The next time the two need to communicate, ATM will be able to handle the connection.

Other services in which ATM plays a key role are looming on the horizon, including wireless ATM and video on demand for the delivery of interactive content such as videoconferencing and television. This leads to what we often refer to as "the great triumvirate:" ATM, SONET or SDH, and broadband services. By combining the powerful switching and multiplexing fabric of ATM with the limitless transport capabilities of SONET or SDH, true broadband services can be achieved, and the idea of creating a network that can be all things to all services can finally be realized.

Figure 3-31
IP switching

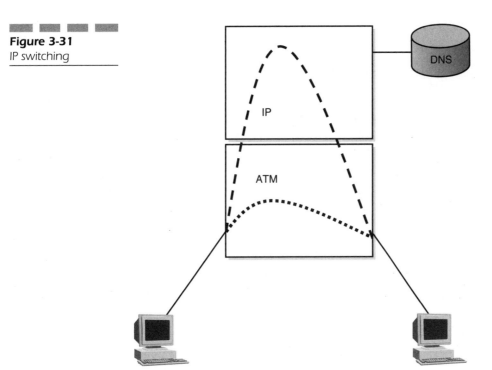

Figure 3-31
IP switching

Optical Networking

Optical networking is often viewed as a point-to-point technology. It has achieved such a position of prominence in the last two years, however, that it qualifies as a transport option in its own right. Furthermore, optical switching is quickly becoming real and optical routing is not far behind. In metro applications, optical has become a major technology player simply because it has the capability to offer unprecedented levels of bandwidth, which minimizes provisioning requirements for the service provider in a geographical area where provisioning and installation activities tend to be prohibitively expensive. With the recent arrival of optical switching and the inclusion of multichannel *Dense Wavelength Division Multiplexing* (DWDM) in metro applications, optical transport has become a central component of cost-effective network services delivery in the metro marketplace.

In this next section, we discuss the development of optical networking, the various technologies that it employs, and the direction it seems to be going.

Beginnings

In 1966, Charles Kao and Charles Hockham at the United Kingdom's Standard Telecommunication Laboratory (now part of Nortel Networks) published a major work, demonstrating that optical fiber could be used to carry information provided its end-to-end signal loss could be kept below 20 decibels per kilometer. Keeping in mind that the decibel scale is logarithmic, 20 decibels of loss means that 99 percent of the light would be lost over each kilometer of distance. Only 1 percent would actually reach the receiver, and that's a 1-kilometer run. Imagine the loss over today's fiber cables that are hundreds of kilometers long, if 20 decibels was the modern performance criterion!

Kao and Hockham proved that metallic impurities in the glass, such as chromium, vanadium, iron, and copper, were the primary cause for such high levels of loss. In response, glass manufacturers rose to the challenge and began to research the creation of ultra-pure products.

In 1970, Peter Schultz, Robert Maurer, and Donald Keck of Corning Glass Works (now Corning Corporation) announced the development of a glass fiber that offered better attenuation than the recognized 20-decibel threshold. Today, fiber manufacturers offer fiber so incredibly pure that 10 percent of the light arrives at a receiver placed 50 kilometers away. Put another way, a fiber with 0.2 decibels of measured loss delivers more than 60 percent of the transmitted light over a distance of 10 kilometers. Remember the windowpane example? Imagine glass so pure that you could see clearly through a window 10 kilometers thick.

Fundamentals of Optical Networking

At their most basic level, optical networks require three fundamental components, as shown in Figure 3-32: a source of light, a medium over which to transport it, and a receiver for the light. Additionally, there may be regenerators, optical amplifiers, and other pieces of equipment in the circuit. We will examine each of these generic components in turn.

Optical Sources

Today the most common sources of light for optical systems are either *light-emitting diodes* (LEDs) or laser diodes. Both are commonly used, although laser diodes have become more common for high-speed data applications

Figure 3-32
The components of a typical optical network. The source originates the signal, the sink receives and interprets it, the amplifiers amplify the optical signal as it passes through them, and the regenerator reconstructs the optical signal by converting it first to electrical then back to optical.

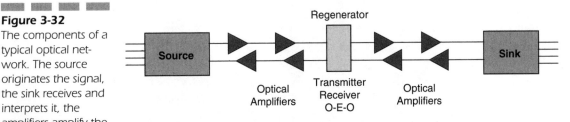

because of their coherent signal. Although lasers have gone through several iterations over the years, including ruby rod and helium neon, semiconductor lasers became the norm shortly after their introduction in the early 1960s because of their low cost and high stability.

Optical Fiber

When Schultz, Maurer, and Keck began their work at Corning to create a low-loss optical fiber, they did so using a newly crafted process called *inside vapor deposition* (IVD). Whereas most glass is manufactured by melting and reshaping silica, IVD deposits various combinations of carefully selected compounds on the inside surface of a silica tube. The tube becomes the cladding of the fiber, and the vapor-deposited compounds become the core. The compounds are typically silicon chloride ($SiCl_4$) and oxygen (O_2), which are reacted under heat to form a soft, sooty deposit of silicon dioxide (SiO_2), as shown in Figure 3-33. In some cases, impurities such as germanium are added at this time to cause various effects in the finished product. In practice, the $SiCl_4$ and O_2 are pumped into the fused silica tube as gases; the tube is heated in a high-temperature lathe, causing the sooty deposit to collect on the inside surface of the tube. The continued heating of the tube causes the soot to fuse into a glass-like substance. Ultimately, once the deposition process is complete, the glass tube is heated fiercely, which causes the tube to collapse, creating what is known in the optical fiber industry as a *preform*. An example of a preform is shown in Figure 3-33.

Figure 3-33
An optical preform,
the first stage in the
creation of optical
fiber

An alternative manufacturing process is called *outside vapor deposition* (OVD). In the OVD process, the soot is deposited on the surface of a rotating ceramic cylinder in two layers. The first layer is the soot that will become the core; the second layer becomes the cladding. Ultimately, the rod and soot are sintered to create a preform. The ceramic is then removed, leaving behind the fused silica that will become the fiber. The next step is to convert the preform into optical fiber.

Drawing the Fiber

To make fiber from a preform, the preform is mounted in a furnace at the top of a tall building called a *drawing tower* (see Figure 3-34). The bottom of the preform is heated until it has the consistency of taffy, at which time the soft glass is drawn down to form a thin fiber. When it strikes the cooler air outside the furnace, the fiber solidifies. Needless to say, the process is carefully managed to ensure that the thickness of the fiber is precise; microscopes are used to verify the geometry of the fiber.

Other stages in the manufacturing process include monitoring processes to check the integrity of the product, a coating process that applies a protective layer, and a take-up stage where the fiber is wound onto reels for later assembly into cables of various types.

Figure 3-34
A drawing tower. The preform is heated to the consistency of taffy on the uppermost floor. It flows to the lower floors on its way to becoming fiber.

Optical Fiber

Fiber is available in different forms to satisfy the varied specification demands of optical transport applications. Some are holdovers from previous generations of optical technology that are still in use and represent the best efforts of technology available from that time. Others represent improvements on the general theme or specialized solutions to specific optical transmission challenges.

Generally speaking, two major types of fiber exist: *multimode,* which is the earliest form of optical fiber and is characterized by a large-diameter central core, short distance capability, and low bandwidth, and *single mode,* which has a narrow core and is capable of greater distance and higher bandwidth. Varieties of each one will be discussed in detail later in the book.

To understand the reason for and philosophy behind the various forms of fiber, it is first necessary to understand the issues that confront transmission engineers who design optical networks.

Optical fiber has a number of advantages over copper. It is lightweight, has enormous bandwidth potential, has significantly higher tensile strength, can support many simultaneous channels, and is immune to electromagnetic interference. It does, however, suffer from several disruptive problems that cannot be discounted. The first of these is *loss* or *attenuation,* the inevitable weakening of the transmitted signal over distance that has a direct analog in the copper world. Attenuation is typically the result of two subproperties, *scattering* and *absorption,* both of which have cumulative effects. The second is *dispersion*, which is the spreading of the transmitted signal and is analogous to noise.

Scattering

Scattering occurs because of impurities or irregularities in the physical makeup of the fiber itself. The best known form of scattering is called *Rayleigh scattering*. It is caused by metal ions in the silica matrix and results in light rays being scattered in various directions.

Rayleigh scattering occurs most commonly around wavelengths of 1000 nm and is responsible for as much as 90 percent of the total attenuation that occurs in modern optical systems. It occurs when the wavelengths of the light being transmitted are roughly the same size as the physical molecular structures within the silica matrix. Thus, short wavelengths are affected by Rayleigh scattering effects far more than long wavelengths. In fact, it is because of Rayleigh scattering that the sky appears to be blue; the shorter (blue) wavelengths of light are scattered more than the longer wavelengths of light.

Absorption

Absorption results from three factors: hydroxyl (OH⁻; water) ions in the silica, impurities in the silica, and incompletely diminished residue from the manufacturing process. These impurities tend to absorb the energy of the transmitted signal and convert it to heat, resulting in an overall weakening of the optical signal. Hydroxyl absorption occurs at 1.25 and 1.39 µm. At 1.7 µm, the silica itself starts to absorb energy because of the natural resonance of silicon dioxide.

Dispersion

As mentioned earlier, dispersion is the optical term for the spreading of the transmitted light pulse as it transits the fiber. It is a bandwidth-limiting phenomenon and comes in two forms: *multimode dispersion* and *chromatic dispersion*. Chromatic dispersion is further subdivided into *material dispersion* and *waveguide dispersion*.

Multimode Dispersion   To understand multimode dispersion, it is first important to understand the concept of a *mode*. Figure 3-35 shows a fiber with a relatively wide core. Because of the width of the core, it allows light rays arriving from the source at a variety of angles (three in this case) to enter the fiber and be transmitted to the receiver. Because of the different paths that each ray, or mode, will take, they will arrive at the receiver at different times, resulting in a dispersed signal.

Now consider the system shown in Figure 3-36. The core is much narrower and only allows a single ray to be sent down the fiber. This results in less end-to-end energy loss and avoids the dispersion problem that occurs in multimode installations.

Chromatic Dispersion   The speed at which an optical signal travels down a fiber is absolutely dependent upon its wavelength. If the signal comprises multiple wavelengths, then the different wavelengths will travel at different speeds, resulting in an overall spreading or smearing of the signal. As discussed earlier, chromatic dispersion comprises two subcategories: *material dispersion* and *waveguide dispersion*.

Material Dispersion   Material dispersion occurs because different wavelengths of light travel at different speeds through an optical fiber. To minimize this particular dispersion phenomenon, two factors must be

Figure 3-35
A multimode fiber. Note the wide core diameter, which allows multiple modes to propagate.

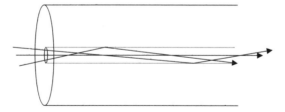

Figure 3-36
A single-mode fiber (SMF). Note the narrow core diameter, which allows only a single mode to propagate.

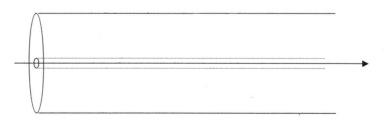

managed. The first of these is the number of wavelengths that make up the transmitted signal. An LED, for example, emits a rather broad range of wavelengths between 30 and 180 nm, whereas a laser emits a much narrower spectrum, typically less than 5 nm. Thus, a laser's output is far less prone to be seriously affected by material dispersion than the signal from an LED.

The second factor that affects the degree of material dispersion is a characteristic called the source signal's *center operating wavelength*. In the vicinity of 850 nm, red, longer wavelengths travel faster than their shorter blue counterparts, but at 1550 nm the situation is the opposite: Blue wavelengths travel faster. Of course, the two eventually meet at a point and share a common minimum dispersion level; it is in the range of 1310 nm, often referred to as the *zero dispersion wavelength*. Clearly, this is an ideal place to transmit data signals, since dispersion effects are minimized here. As we will see later, however, other factors crop up that make this a less desirable transmission window than it appears. Material dispersion is a particularly vexing problem in SMFs.

Waveguide Dispersion Because the core and cladding of a fiber have slightly different indices of refraction, the light that travels in the core moves slightly slower than the light that escapes into and travels in the cladding. This results in a dispersion effect that can be corrected by transmitting at specific wavelengths where material and waveguide dispersion actually occur at a minimum.

Putting It All Together

So what does all of this have to do with the high-speed transmission of voice, video, and data? A lot, as it turns out. Understanding where attenuation and dispersion problems occur helps optical design engineers determine the best wavelengths at which to transmit information, taking into account distance, the type of fiber, and other factors that can potentially affect the integrity of the transmitted signal.

Consider the graph shown in Figure 3-37. It depicts the optical transmission domain as well as the areas where problems arise. Attenuation (dB/km) is shown on the Y-Axis, and wavelength (nm) is shown on the X-Axis.

First of all, note that there are four *transmission windows* in the diagram. The first one is at approximately 850 nm, the second at 1310 nm, a

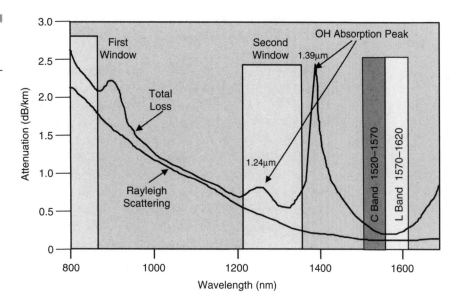

Figure 3-37
The optical
transmission domain

third at 1550 nm, and a fourth at 1625 nm, the last two labeled C and L band, respectively. The 850 nm band was the first to be used because of its adherence to the wavelength at which the original LED technology operated. The second window at 1310 nm enjoys low dispersion; this is where dispersion effects are minimized. 1550 nm, the so-called C band, has emerged as the ideal wavelength at which to operate long-haul systems and systems upon which DWDM has been deployed because (1) loss is minimized in this region, and (2) dispersion minimums can be shifted here. The relatively new L band has enjoyed some early success as the next effective operating window.

Notice also that Rayleigh scattering is shown to occur at or around 1000 nm, while hydroxyl absorption by water occurs at 1240 and 1390 nm. Needless to say, network designers would be well served to avoid transmitting at any of the points on the graph where Rayleigh scattering, high degrees of loss, or hydroxyl absorption have the greatest degree of impact. Notice also that dispersion, shown by the lower line, is at a minimum point in the second window, while loss, shown by the upper line, drops to a minimum point in the third window. In fact, dispersion is minimized in traditional SMF at 1310 nm, while loss is at minimums at 1550 nm. So the obvious question becomes this: Which one do you want to minimize—loss or dispersion?

Luckily, this choice no longer has to be made. Today, *dispersion-shifted fiber* (DSF) has become common. By modifying the manufacturing process, engineers can shift the point at which minimum dispersion occurs from 1310 nm to 1550 nm, causing it to coincide with the minimum loss point such that loss and dispersion occur at the same wavelength.

Unfortunately, although this fixed one problem, it created a new and potentially serious alternative problem. DWDM has become a mainstay technology for multiplying the available bandwidth in optical systems. When DWDM is deployed over DSF, serious nonlinearities occur at the zero dispersion point and effectively destroy the DWDM signal. Think about it: DWDM relies on the capability to "channelize" the available bandwidth of the optical infrastructure and maintain some degree of separation between the channels. If dispersion is minimized in the 1559 nm window, then the channels will effectively overlay each other in DWDM systems.

Specifically, a problem called *four-wave mixing* creates "sidebands" that interfere with the DWDM channels, destroying their integrity. In response, fiber manufacturers have created *nonzero dispersion-shifted fiber* (NZDSF) that lowers the dispersion point to near zero, making it occur just outside of the 1550 nm window. This eliminates the nonlinear four-wave mixing problem.

Fiber Nonlinearities

As mentioned earlier, the classic business quote, "in its success lie the seeds of its own destruction," is imminently applicable to the optical networking world. As the marketplace clamors for longer transmission distances with minimal amplification, more wavelengths per fiber, higher bit rates, and increased signal power, a rather ugly collection of transmission impairments, known as *fiber nonlinearities,* rises to challenge the attempts to meet such demands. These impairments go far beyond the simple concerns brought about by loss and dispersion; they represent a significant performance barrier.

The special relationship that exists between transmission power and the refractive index of the medium gives rise to four service-affecting optical nonlinearities: *self-phase modulation* (SPM), *cross-phase modulation* (XPM), *four-wave mixing*, and intermodulation.

Self-Phase Modulation (SPM)

When SPM occurs, chromatic dispersion kicks in to create something of a technological double-whammy. As the light pulse moves down the fiber, its leading edge increases the refractive index of the core, which causes a shift toward the longer-wavelength, blue end of the spectrum. The trailing edge, on the other hand, decreases the refractive index of the core, causing a shift toward the shorter-wavelength, red end of the spectrum. This causes an overall spreading or smearing of the transmitted signal, a phenomenon known as *chirp*. It occurs in fiber systems that transmit a single pulse down the fiber and is proportional to the amount of chromatic dispersion in the fiber: the more chromatic dispersion, the more SPM. It is counteracted with the use of large, effective area fibers.

Cross-Phase Modulation (XPM)

When multipleXPM optical signals travel down the same fiber core, they both change the refractive index in direct proportion to their individual power levels. If the signals happen to cross, they will distort each other. Although XPM is similar to SPM, one significant difference exists: Although SPM is directly affected by chromatic dispersion, XPM is only minimally affected by it. Large, effective area fibers can reduce the impact of XPM.

Four-Wave Mixing

Four-wave mixing is the most serious of the power/refractive index-induced nonlinearities today because it has a catastrophic effect on DWDM-enhanced systems. Because the refractive index of fiber is nonlinear, and because multiple optical signals travel down the fiber in DWDM systems, a phenomenon known as *third-order distortion* can occur that seriously affects multichannel transmission systems. Third-order distortion causes harmonics to be created in large numbers that have the annoying habit of occurring where the actual signals are, resulting in their obliteration.

Four-wave mixing is directly related to DWDM. In DWDM fiber systems, multiple simultaneous optical signals are transmitted across an optical span. They are separated on an ITU-blessed standard transmission grid by as much as 100 GHz (although most manufacturers today have reduced that to 50 GHz or better). This separation ensures that they do not interfere with each other.

Consider now the effect of DSF on DWDM systems. In DSF, signal transmission is moved to the 1550 nm band to ensure that dispersion and loss are both minimized within the same window. However, minimal dispersion has a rather severe unintended consequence when it occurs in concert with

DWDM: Because it reduces dispersion to near zero, it also prevents multi-channel systems from existing since it does not enable proper channel spacing. Four-wave mixing, then, becomes a serious problem.

Several things can reduce the impact of four-wave mixing. As the dispersion in the fiber drops, the degree of four-wave mixing increases dramatically. In fact, it is worse at the zero-dispersion point. Thus, the intentional inclusion of a small amount of chromatic dispersion actually helps to reduce the effects of four-wave mixing. For this reason, fiber manufacturers sell NZDSF, which moves the dispersion point to a point near the zero point, thus ensuring that a small amount of dispersion creeps in to protect against four-wave mixing problems.

Another factor that can minimize the impact of four-wave mixing is to widen the spacing between DWDM channels. This, of course, reduces the efficiency of the fiber by reducing the total number of available channels and is therefore not a popular solution, particularly since the trend in the industry is to move toward more narrow channel spacing as a way to increase the total number of available channels. Already several vendors have announced spacing as narrow as 5 GHz. Finally, large effective area fibers tend to suffer less from the effects of four-wave mixing.

Intermodulation Effects

In the same way that XPM results from interference between multiple simultaneous signals, intermodulation causes secondary frequencies to be created that are cross-products of the original signals being transmitted. Large, effective area fibers can alleviate the symptoms of intermodulation.

Scattering Problems

Scattering within the silica matrix causes the second major impairment phenomenon. Two significant nonlinearities result: *Stimulated Brillouin Scattering* (SBS) and *Stimulated Raman Scattering* (SRS).

Stimulated Brillouin Scattering (SBS)   SBS is a power-related phenomenon. As long as the power level of a transmitted optical signal remains below a certain threshold, usually on the order of three milliwatts, SBS is not a problem. The threshold is directly proportional to the fiber's effective area, and because dispersion-shifted fibers typically have smaller effective areas, they have lower thresholds. The threshold is also proportional to the width of the originating laser pulse: As the pulse gets wider, the threshold goes up. Thus, steps are often taken through a variety of techniques to artificially broaden the laser pulse.

SBS is caused by the interaction of the optical signal moving down the fiber with the acoustic vibration of the silica matrix that makes up the fiber. As the silica matrix resonates, it causes some of the signal to be reflected back toward the source of the signal, resulting in noise, signal degradation, and a reduction of the overall bit rate in the system. As the power of the signal increases beyond the threshold, more of the signal is reflected, resulting in a multiplication of the initial problem.

It is interesting to note that actually two forms of Brillouin scattering exist. When electric fields that oscillate in time within an optical fiber interact with the natural acoustic resonance of the fiber material itself, the result is a tendency to backscatter light as it passes through the material, or Brillouin scattering. If, however, the electric fields are caused by the optical signal itself, the signal is seen to cause the phenomenon, which is SBS.

To summarize: Because of backscattering, SBS reduces the amount of light that actually reaches the receiver and causes noise impairments. The problem increases quickly above the threshold and has a more deleterious impact on longer wavelengths of light. One additional fact: In-line optical amplifiers such as *erbium-doped fiber amplifiers* (EDFAs) add to the problem significantly. If four optical amplifiers are located along an optical span, the threshold will drop by a factor of four. Solutions to SBS include the use of wider-pulse lasers and larger effective area fibers.

Stimulated Raman Scattering (SRS)   SRS is something of a power-based crosstalk problem. In SRS, high-power, short-wavelength channels tend to bleed power into longer-wavelength, lower-power channels. SRS occurs when a light pulse moving down the fiber interacts with the crystalline matrix of the silica, causing the light to (1) be backscattered and (2) shift the wavelength of the pulse slightly. Whereas SBS is a backward-scattering phenomenon, SRS is a two-way phenomenon, causing both backscattering and a wavelength shift. The result is crosstalk between adjacent channels.

The good news is that SRS occurs at a much higher power level, close to a watt. Furthermore, it can be effectively reduced through the use of large, effective area fibers.

Optical Amplification

As long as we are on the subject of Raman scattering, we should introduce the concept of optical amplification. This may seem like a bit of a non sequitur, but it really isn't. True optical amplification actually uses a form of Raman scattering to amplify the transmitted signal.

Traditional Amplification and Regeneration Techniques

In a traditional metallic analog environment, transmitted signals tend to weaken over distance. To overcome this problem, amplifiers are placed in the circuit periodically to raise the power level of the signal. This technique has a problem, however: In addition to amplifying the signal, amplifiers also amplify whatever cumulative noise has been picked up by the signal during its trip across the network. Over time, it becomes difficult for a receiver to discriminate between the actual signal and the noise embedded in the signal. Extraordinarily complex recovery mechanisms are required to discriminate between optical wheat and noise chaff.

In digital systems, *regenerators* are used to not only amplify the signal, but to also remove any extraneous noise that has been picked up along the way. Thus, digital regeneration is a far more effective signal recovery methodology than simple amplification.

Even though signals propagate significantly farther in optical fiber than they do in copper facilities, they are still eventually attenuated to the point that they must be regenerated. In a traditional installation, the optical signal is received by a receiver circuit, converted to its electrical analog, regenerated, converted back to an optical signal, and transmitted onward over the next fiber segment. This *optical-to-electrical-to-optical* (O-E-O) conversion process is costly, complex, and time consuming. However, it is proving to be far less necessary as an amplification technique than it used to be because of true optical amplification that has recently become commercially feasible. Please note that optical amplifiers *do not* regenerate signals; they merely amplify. Regenerators are still required, albeit far less frequently.

Optical amplifiers represent one of the technological leading edges of data networking. Instead of the O-E-O process, optical amplifiers receive the optical signal, amplify it as an optical signal, and then retransmit it as an optical signal; no electrical conversion is required. Like their electrical counterparts, however, they also amplify the noise; at some point signal regeneration is required.

Optical Amplifiers: How They Work

It was only a matter of time before all-optical amplifiers became a reality. It makes intuitively clear sense that a solution that eliminates the electrical portion of the O-E-O process would be a good one. Optical amplification is that solution.

You will recall that SRS is a fiber nonlinearity that is characterized by high-energy channels pumping power into low-energy channels. What if that phenomenon could be harnessed as a way to amplify optical signals that have weakened over distance?

Optical amplifiers are actually rather simple devices that, as a result, tend to be extremely reliable. The optical amplifier comprises the following: an input fiber carrying the weakened signal that is to be amplified, a pair of optical isolators, a coil of doped fiber, a pump laser, and the output fiber that now carries the amplified signal. A functional diagram of an optical amplifier is shown in Figure 3-38.

The coil of doped fiber lies at the heart of the optical amplifier's functionality. Doping is simply the process of embedding some kind of functional impurity in the silica matrix of the fiber when it is manufactured. In optical amplifiers, this impurity is more often than not an element called *erbium*. Its role will become clear in just a moment.

The pump laser shown in the upper-left corner of Figure 3-38 generates a light signal at a particular frequency (often 980 nm) in the opposite direction that the actual transmitted signal flows. As it turns out, erbium

Figure 3-38
In this EDFA, as the weakened signal flows into the segment of doped fiber, the pump laser excites the erbium atoms embedded in the silicon matrix of the fiber. They release photons at the exact wavelength required to amplify the incoming signal.

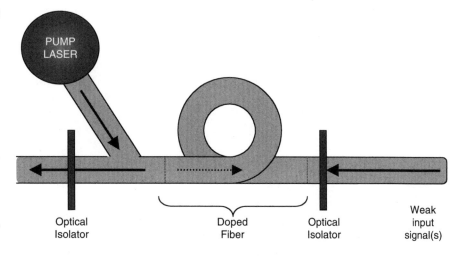

PUMP LASER

Optical Isolator

Doped Fiber

Optical Isolator

Weak input signal(s)

becomes atomically excited when it is struck by light at that wavelength. When an atom is excited by pumped energy, it jumps to a higher energy level (those of you recovering physicists will remember classroom discussions about orbital levels: $1S^1$, $1S^2$, $2S^1$, $2S^2$, $2P^6$, and so on). The atom then falls back down, during which it gives off a photon at a certain wavelength. When erbium is excited by light at 980 nm, it emits photons within the 1550 nm region, coincidentally the wavelength at which multichannel optical systems operate. So when the weak, transmitted signal reaches the coil of erbium-doped fiber, the erbium atoms, now excited by the energy from the pump laser, bleed power into the weak signal at precisely the right wavelength, causing a generalized amplification of the transmitted signal. The optical isolators serve to prevent errant light from backscattering into the system, creating noise.

EDFAs amplify anything, including the noise that the signal may have picked up. Thus, regeneration will still be needed at some point along the path of long-haul systems, although far less frequently than in traditional copper systems. Most manufacturers of optical systems publish recommended span engineering specifications that help service providers and network designers take such concerns into account as they design each transmission facility.

Optical Receivers

So far, we have discussed the sources of light, including LEDs and laser diodes, as well as the various flavors of optical fiber and the problems they encounter as transmission media. Now we turn our attention to the devices that receive the transmitted signal.

The receive devices used in optical networks have a single responsibility: to capture the transmitted optical signal and convert it into an electrical signal that can then be processed by the end equipment. Various stages of amplification may ensure that the signal is strong enough to be acted upon, and demodulation circuitry may recreate the originally transmitted electronic signal.

Photodetector Types

Although many different types of photosensitive devices are available, two are used most often as photodetectors in modern networks: *positive-intrinsic-negative* (PIN) photodiodes and *avalanche photodiodes* (APDs).

Optical Fiber

Fiber has evolved over the years in a variety of ways to accommodate both the changing requirements of the customer community and the technological challenges that emerged as the demand for bandwidth climbed precipitously. These changes came in various forms of fiber that presented different behavior characteristics to the market.

Modes: An Analogy

The concept of modes is sometimes difficult to understand, so let me pass along an analogy that will help. Imagine a shopping mall that has a wide, central area that all the shops open onto. An announcement comes over the PA system informing people that "The mall is now closed; please make your way to the exit." Shoppers begin to make their way to the doors, but some wander from store to store, windowshopping along the way, while others take a relatively straight route to the exit. The result is that some shoppers take longer than others to exit the mall because there are different modes.

Now consider a mall that has a single, very narrow corridor that is only as wide as a person's shoulders. Now when the announcement comes, everyone heads for the exit, but they must form a single-file line and head out in an orderly fashion. If you understand the difference between these two examples, you understand single versus multimode fiber. The first example represents multimode; the second represents single mode.

Multimode Fiber The first of these is *multimode fiber*, which arrived in a variety of different forms. Multimode fiber bears that name because it enables more than a single mode or ray of light to be carried through the fiber simultaneously because of the relatively wide core diameter that characterizes the fiber. Although the dispersion that potentially results from this phenomenon can be a problem, multimode fiber has its advantages. For one thing, it is far easier to couple the relatively wide and forgiving end of a multimode fiber to a light source than that of the much narrower SMF. It is also significantly less expensive to manufacture (and purchase), and it relies on LEDs and inexpensive receivers rather than the more expensive laser diodes and ultra-sensitive receiver devices. However, advancements in technology have caused the use of multimode fiber to fall out of favor; single mode is far more commonly used today.

Multimode fiber is manufactured in two forms: *step-index fiber* and *graded-index fiber*. In step-index fiber, the index of refraction of the core is slightly higher than the index of refraction of the cladding. Remember that the higher the refractive index, the slower the signal travels through the medium. Thus, in step-index fiber, any light that escapes into the cladding because it enters the core at too oblique an angle will actually travel slightly faster in the cladding (assuming it does not escape altogether) than it would if it traveled in the core. Of course, any rays that are reflected repeatedly as they traverse the core also take longer to reach the receiver, resulting in a dispersed signal that causes problems for the receiver at the other end. Clearly, this phenomenon is undesirable; for that reason, graded-index fiber was developed.

In graded-index fiber, the refractive index of the core actually decreases from the center of the fiber outward. In other words, the refractive index at the center of the core is higher than the refractive index at the edge of the core. The result of this rather clever design is that as light enters the core at multiple angles and travels from the center of the core outward, it is actually accelerated at the edge and slowed down near the center, causing most of the light to arrive at roughly the same time. Thus, graded-index fiber helps to overcome the dispersion problems associated with step-index multimode fiber. Light that enters this type of fiber does not travel in a straight line but rather follows a parabolic path, with all rays arriving at the receiver at more or less the same time.

Graded-index fiber was commonly used in telecommunications applications until the late 1980s. Even though graded-index fiber is significantly better than step-index fiber, it is still multimode fiber and does not eliminate the problems inherent in being multimode. Thus was born the next generation of optical fiber: single-mode.

Single-Mode Fiber (SMF)   An interesting mental conundrum crops up with the introduction of SMF. The core of SMF is significantly narrower than the core of multimode fiber. Because it is narrower, it would seem that its capability to carry information would be reduced due to a limited light-gathering capability. This, of course, is not the case. As its name implies, it enables a single mode or ray of light to propagate down the fiber core, thus eliminating the intermodal dispersion problems that plague multimode fibers. In reality, SMF is a stepped-index design, because the core's refractive index is slightly higher than that of the cladding. It has become the de facto standard for optical transmission systems and takes on many forms depending on the specific application within which it will be used.

Most SMF has an extremely narrow core diameter on the order of 7 to 9 microns, and a cladding diameter of 125 microns. The advantage of this design is that it only enables a single mode to propagate; the downside, however, is the difficulty involved in working with it. The core must be coupled directly to the light source and the receiver in order to make the system as effective as possible. Given that the core is approximately one-sixth the diameter of a human hair, the mechanical process through which this coupling takes place becomes Herculean.

Since its introduction in the early 1980s, SMF has undergone a series of evolutionary phases in concert with the changing demands of the bandwidth marketplace. The first variety of SMF to enter the market was called *nondispersion-shifted fiber* (NDSF). Designed to operate in the 1310 nm second window, dispersion in these fibers was close to zero at that wavelength. As a result, it offered high bandwidth and low dispersion. Unfortunately, it was soon the victim of its own success. As demand for high-bandwidth transport grew, a third window was created at 1550 nm for SMF transmission. It provided attenuation levels that were less than half those measured at 1310 nm, but unfortunately it was plagued with significant dispersion. Since the bulk of all installed fiber was NDSF, the only solution available to transmission designers was to narrow the linewidth of the lasers employed in these systems and to make them more powerful. Unfortunately, increasing the power and reducing the laser linewidth is expensive, so another solution emerged.

Dispersion-Shifted Fiber (DSF)   One solution that emerged was DSF. With DSF, the minimum dispersion point is mechanically shifted from 1310 nm to 1550 nm by modifying the design of the actual fiber so that waveguide dispersion is increased. The reader will recall that waveguide dispersion is a form of chromatic dispersion that occurs because the light travels at different speeds in the core and cladding.

One technique for building DSF (sometimes called ZDSF) is to actually build a fiber of multiple layers. In this design, the core has the highest index of refraction and changes gradually from the center outward until it equals the refractive index of the outer cladding. The inner core is surrounded by an inner cladding layer, which is in turn surrounded by an outer core. This design works well for single wavelength systems, but it experiences serious signal degradation when multiple wavelengths are transmitted, such as when used with DWDM systems. Four-wave mixing, described earlier, becomes a serious impediment to clean transmission in these systems. Given that multiple wavelength systems are quickly becoming the norm today, the single wavelength limit is a show-stopper. The result is a relatively simple and elegant set of solutions.

The second technique is to eliminate or at least substantially reduce the absorption peaks in the fiber performance graph so that the second and third transmission windows merge into a single, larger window. This allows for the creation of the fourth window described earlier that operates between 1565 and 1625 nm, the so-called L-Band.

Finally, the third solution comes with the development of NZDSF. NZDSF shifts the minimum dispersion point so that it is *close* to the zero point, but not actually *at* it. This prevents the nonlinear problems that occur at the zero point to be avoided because it introduces a small amount of chromatic dispersion.

Summary

In this section, we have examined the history of optical technology and the technology itself, focusing on the three key components within an optical network: the light emitter, the transport medium, and the receiver. We also discussed the various forms of transmission impairment that can occur in optical systems and the steps that have been taken to overcome them.

The result of all this is that optical fiber, once heralded as a near-technological miracle because it only lost 99 percent of its signal strength *when transmitted over an entire kilometer,* has become the standard medium for the transmission of high-bandwidth signals over great distances. Optical amplification now serves as an augmentation to traditional regenerated systems, allowing for the elimination of the optical-to-electrical conversion that must take place in copper systems. The result of all this is an extremely efficient transmission system that has the capability to play a role in virtually any network design in existence today.

Dense Wavelength Division Multiplexing (DWDM)

When high-speed transport systems such as SONET and SDH were first introduced, the bandwidth that they made possible was unheard of. The early systems that operated at OC-3/STM-1 levels (155.52 Mbps) provided volumes of bandwidth that were almost unimaginable. As the technology advanced to higher levels, the market followed Say's Law, creating demand for the ever more available volumes of bandwidth. There were limits, however.

Today OC-48/STM-16 (2.5 Gbps) is extremely popular, but OC-192/STM-64 (10 Gbps) represents the practical upper limit of SONET's and SDH's transmission capabilities given the limitations of existing TDM technology. The alternative is to simply multiply the channel count—and that's where WDM comes into play.

WDM is really nothing more than frequency division multiplexing, albeit at very high frequencies. The ITU has standardized a channel separation grid that centers around 193.1 THz, ranging from 191.1 THz to 196.5 THz. Channels on the grid are technically separated by 100 GHz, but many industry players today are using 50 GHz separation.

The majority of WDM systems operate in the C band (third window, 1550 nm), which enables the close placement of channels and the reliance on EDFAs to improve signal strength. Older systems, which spaced the channels 200 GHz (1.6 nm) apart, were referred to simply as WDM systems; the newer systems are referred to as *Dense* WDM systems because of their tighter channel spacing. Modern systems routinely pack 40 to 10 Gbps channels across a single fiber, for an aggregate bit rate of 400 Gbps.

How DWDM Works

As Figure 3-39 illustrates, a WDM system consists of multiple input lasers, an ingress multiplexer, a transport fiber, an egress multiplexer, and, of course, customer-receiving devices. If the system has eight channels such as the one shown in the diagram, it has eight lasers and eight receivers. The channels are separated by 100 GHz to avoid fiber nonlinearities, or are closer if the system supports the 50 GHz spacing. Each channel, sometimes referred to as a lambda (λ, the Greek letter and universal symbol used to represent wavelength), is individually modulated, and ideally the signal strengths of the channels should be close to one another. Generally speaking, this is not a problem, because in DWDM systems the channels are

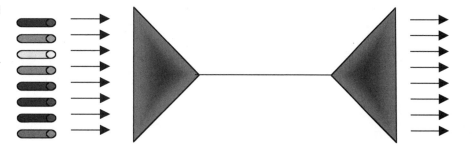

Figure 3-39
DWDM channel separation

closely spaced and therefore do not experience significant attenuation variation from channel to channel.

Operators of DWDM-equipped networks face a significant maintenance issue. Consider a 16-channel DWDM system, a common arrangement in metro installations. This system has 16 lasers, one for each channel, which means that the service provider must maintain 16 spare lasers in case of a laser failure. The latest effort underway is the deployment of tunable lasers, which enable the laser to be tuned to any output wavelength, thus reducing the volume of spares that must be maintained and by extension, the cost.

So what do we find in a typical WDM system? A variety of components are used, as follows:

- *Multiplexers*, which combine multiple optical signals for transport across a single fiber
- *Demultiplexers*, which disassemble the aggregate signal so that each signal component can be delivered to the appropriate optical receiver (PIN or APD)
- *Active or passive switches or routers*, which direct each signal component in a variety of directions
- *Filters*, which serve to provide wavelength selection
- *Optical ADMs*, which give the service provider the ability to pick up and drop off individual wavelength components at intermediate locations throughout the network

Together these components make up the heart of the typical high-bandwidth optical network. So why is DWDM so important? Because of the cost differential that exists between a DWDM-enhanced network and a traditional network. Expanding network capacity today by putting more fiber in the ground costs, on average, about $70K per mile. To add the same bandwidth using DWDM by changing the endpoint electronics costs roughly one-sixth that amount. Clearly, the WDM solution has a greater financial incentive.

Coarse Wavelength Division Multiplexing (CWDM)

As the use of optical technology has grown in metro networks, service providers have found that technologies such as DWDM enable them to deploy relatively low-cost, high-bandwidth solutions without stranding an entire fiber with a single customer. At the same time, because these are metro solutions, they are not hindered by the distance limitations that plague long-haul technologies. Thus, relatively low-cost solutions such as *coarse wavelength division multiplexing* (CWDM) can offer a lower channel count without the cost, challenge, and complexity of amplification and tight channel spacing. Typical CWDM solutions offer 16- or 32-channel systems, typically adequate for the relatively limited requirements of many metro customers.

Optical Switching and Routing

DWDM facilitates the transport of massive volumes of data from a source to a destination. Once the data arrives at the destination, however, it must be terminated and redirected to its final destination on a lambda-by-lambda basis. This is done with switching and routing technologies.

Switching Versus Routing: What's the Difference?

A review of these two fundamental technologies is probably in order. The two terms are often used interchangeably, and a never-ending argument is under way about the differences between the two.

The answer lies in the lower layers of the now-famous OSI Protocol Model (see Figure 3-40). You will recall that OSI is a conceptual model used to study the step-by-step process of transmitting data through a network. It comprises seven layers, the lower three of which define the domain of the typical service provider. These layers, starting with the lowest in the seven-layer stack, are the physical layer (layer one), the data link layer (layer two), and the network layer (layer three). Layer one is responsible for defining the standards and protocols that govern the physical transmission of bits across a medium. SONET and SDH are both physical-layer standards.

Switching, which lies at layer 2 (the data link layer) of OSI, is usually responsible for establishing connectivity within a single network. It is a relatively low-intelligence function and is therefore accomplished quite quickly. Such technologies as ATM and Frame Relay; wireless access

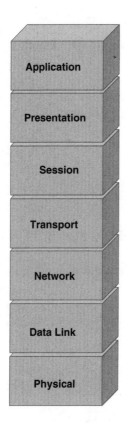

Figure 3-40
The seven-layer OSI
Reference Model

Application

Presentation

Session

Transport

Network

Data Link

Physical

technologies such as *Frequency Division Multiple Access* (FDMA), TDMA, and CDMA; and LAN access control protocols (CSMA/CD and token passing) are found at this layer.

Routing, on the other hand, is a layer 3 (network layer) function. It operates at a higher, more complex level of functionality and is therefore more complex. Routing concerns itself with the movement of traffic between subnetworks and therefore complements the efforts of the switching layer. ATM, Frame Relay, LAN protocols, and the PSTN are switching protocols, while IP, the *Routing Information Protocol* (RIP), *Open Shortest Path First* (OSPF), and *Internetwork Packet Exchange* (IPX) are routing protocols.

Switching in the Optical Domain

The principal form of optical switching is really nothing more than a sophisticated digital cross-connect system. In the early days of data networking, dedicated facilities were created by manually patching the end points of a circuit at a patch panel, thus creating a complete four-wire circuit. Beginning in the 1980s, digital cross-connect devices such as AT&T's *Digital Access and Cross-Connect* (DACS) became common, replacing the time-consuming, expensive, and error-prone manual process. The digital cross-connect is really a simple switch, designed to establish "long-term temporary" circuits quickly, accurately, and inexpensively.

Enter optical. Traditional cross-connect systems worked fine in the optical domain, provided no problem occurred going through the O-E-O conversion process. This, however, was one of the aspects of optical networking that network designers wanted to eradicate from their functional requirements. Thus was born the optical cross-connect switch.

The first of these to arrive on the scene was Lucent Technologies' LambdaRouter. Based on a switching technology called *Micro Electrical Mechanical System* (MEMS), the LambdaRouter was the world's first all-optical cross-connect device.

MEMS relies on micro-mirrors, an array of which is shown in Figure 3-41. The mirrors can be configured at various angles to ensure that an incoming lambda strikes one mirror, reflects off a fixed mirrored surface, strikes another movable mirror, and is then reflected out an egress fiber. The LambdaRouter and other devices like it are now commercially deployed and offer speed, a relatively small footprint, bit rate and protocol transparency, nonblocking architecture, and highly developed database management. Fundamentally, these devices are very high-speed, high-capacity switches or cross-connect devices. They are not routers, because they do not perform layer-three functions. They will, however; all major manufacturers have announced plans to incorporate a layer-three routing function in their devices.

Figure 3-41
MEMS mirror array
(Source: Lucent
Technologies)

One problem that quickly reared its head with optical switching was management. When optical switches first arrived on the networking scene, the optical-optical-optical architecture prevented network managers from monitoring the signal strength and other parameters that they were accustomed to monitoring in copper networks, and which were trivially easy to deal with. As a result, some providers allowed their engineers to convert the optical input to an optical switch into a metallic signal so that they could monitor it. Then they converted it back to optical before entering the switch. Clearly, this added a layer of complexity that the optical switching evolution was designed to avoid. Today, more than 30 vendors sell optical management packages to obviate the need for this additional conversion.

Optical networking, then, is a major component of the modern transport network. Together with switching technologies like Frame Relay and ATM, and physical-layer standards like SONET, SDH, and copper-based transport technologies, the current network can handle the transport requirements of *any* traffic type while overlaying QoS at the same time.

SONET brings with it a subset of advantages that makes it stand above competitive technologies. These include midspan meet, improved OAM&P, support for multipoint circuit configurations, nonintrusive facility monitoring, and the capability to deploy a variety of new services. We will examine each of these in turn.

SONET Advantages: Midspan Meet

Because of the monopoly nature of early networks, interoperability was a laughable dream. Following the divestiture of AT&T, however, and the realization of equal access, the need for interoperability standards became a matter of some priority. Driven largely by MCI, the newly competitive telecommunications industry fought hard for standards that would allow different vendors' optical multiplexing equipment to interoperate. This interoperability came to be known as midspan meet, SONET's greatest contribution to the evolving industry.

Improved OAM&P

Improved OAM&P is without question one of the most important contributions that SONET brings to the networking table. Element and network monitoring, management, and maintenance have always been something of a catch-as-catch-can effort because of the complexity and diversity of elements in a typical service provider's network. SONET overhead includes error-checking capabilities, bytes for network survivability, and a diverse set of clearly defined management messages.

Multipoint Circuit Support

When SONET was first deployed on the network, the bulk of the traffic it carried derived from point-to-point circuits such as T1 and DS-3 facilities. With SONET came the ability to hub the traffic, a process that combines the best of cross-connection and multiplexing to perform a capability known as *groom and fill*. This means that aggregated traffic from multiple sources can be transported to a hub, managed as individual components, and redirected out any of several outbound paths without having to completely disassemble the aggregate payload. Prior to SONET, this process required a pair of back-to-back multiplexers, sometimes called an M13 (for "multiplexer that interfaces between DS-1 and DS-3"). This capability, combined with SONET's discreet and highly capable management features, results in a wonderfully manageable system of network bandwidth control.

Nonintrusive Monitoring

SONET overhead bytes are embedded in the frame structure, meaning that they are universally transported alongside the customer's payload. Thus, tight and granular control over the entire network can be realized, leading to more efficient network management and the capability to deploy services on an as-needed basis.

New Services

SONET bandwidth is imminently scalable, meaning that the capability to provision additional bandwidth for customers that require it on an as-needed basis becomes real. As applications evolve to incorporate more and more multimedia content and to therefore require greater volumes of bandwidth, SONET offers it by the bucketload. Interfaces between SONET and Gigabit Ethernet are being written already, while interfaces to ATM and other high-speed switching architectures have been in existence for some time already.

SONET Evolution

SONET was initially designed to provide multiplexed point-to-point transport. However, as its capabilities became better understood and networks became "mission critical," its deployment became more innovative, and soon it was deployed in ring architectures. These rings represent one of the most commonly deployed network topologies. For the moment, however, let's examine a point-to-point deployment. As it turns out, rings don't differ all that much.

If we consider the structure and function of the typical point-to-point circuit, we find a variety of devices and functional regions. The components include the following:

- *End devices*, multiplexers in this case, which provide the point of entry for traffic originating in the customer's equipment and seeking transport across the network
- A *full-duplex circuit*, which provides simultaneous two-way transmission between the network components
- A *series of repeaters or regenerators*, responsible for periodically reframing and regenerating the digital signal
- *One or more intermediate multiplexers*, which serve as nothing more than pass-through devices

When non-SONET traffic is transmitted into a SONET network, it is packaged for transport through a step-by-step, quasi-hierarchical process that attempts to make reasonably good use of the available network bandwidth and ensure that receiving devices can interpret the data when it arrives. The intermediate devices, including multiplexers and repeaters, also play a role in guaranteeing traffic integrity, and to that end the SONET standards divide the network into three regions: path, line, and section. To understand the differences between the three, let's follow a typical transmission of a DS-3, probably carrying 28 T1s, from its origination point to the destination.

When the DS-3 first enters the network, the ingress SONET multiplexer packages it by wrapping it in a collection of additional information, called *path overhead*, which is unique to the transported data. For example, it attaches information that identifies the original source of the DS-3, so that it can be traced in the event of network transmission problems, a bit-error control byte, information about how the DS-3 is actually mapped into the payload transport area (and unique to the payload type), an area for network performance and management information, and a number of other informational components that have to do with the end-to-end transmission of the unit of data.

The packaged information, now known as a *payload,* is inserted into a SONET frame, and at that point another layer of control and management information is added, called *line overhead.* Line overhead is responsible for managing the movement of the payload from multiplexer to multiplexer. To do this, it adds a set of bytes that enables receiving devices to find the payload inside the SONET frame. As you will learn a bit later, the payload can occasionally wander around inside the frame due to the vagaries of the network. These bytes enable the system to track that movement.

In addition to these tracking bytes, the line overhead includes bytes that monitor the integrity of the network and have the capability to switch to a backup transmission span if a failure in the primary span occurs. Line overhead also includes another bit-error checking byte, a robust channel for transporting network management information, and a voice communications channel that enables technicians at either end of a line to plug in with a handset (sometimes called a butt-in, or buttinski) and communicate while troubleshooting.

The final step in the process is to add a layer of overhead that enables the intermediate repeaters to find the beginning of and synchronize a received frame. This overhead, called the *section overhead,* contains a unique initial framing pattern at the beginning of the frame, an identifier for the payload signal being carried, and another bit-error check. It also contains a voice communications channel and a dedicated channel for network management information, similar to but smaller than the one identified in the line overhead.

The result of all this overhead, much of which seems like overkill (and in many peoples' minds *is*), is that the transmission of a SONET frame containing user data can be identified and managed with tremendous granularity from the source all the way to the destination.

So, to summarize, the hard little kernel of DS-3 traffic is gradually surrounded by three layers of overhead information, as shown in Figure 3-42, that help it achieve its goal of successfully traveling across the network. The section overhead is used at every device the signal passes through, including multiplexers and repeaters, the line overhead is only used between multiplexers, and the information contained in the path overhead is only used by the source and destination multiplexers. The intermediate multiplexers

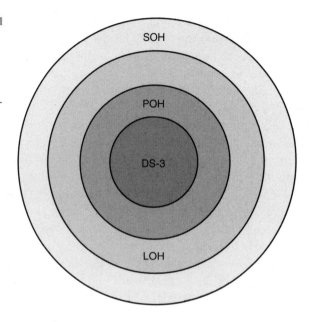

don't care about the specific nature of the payload, because they don't have
to terminate or interpret it.

The SONET Frame

Keep in mind once again that we are doing nothing more complicated than
building a T1 frame with an attitude. Recall that the T1 frame is comprised
of 24 eight-bit channels (samples from each of 24 incoming data streams)
plus a single bit of overhead. In SONET, we have a similar construct—a lot
more channel capacity, and a lot more overhead, but the same functional
concept.

The fundamental SONET frame is shown in Figure 3-43 and is known as
a *Synchronous Transport Signal, Level One* (STS-1). It is 9 bytes tall and 90
bytes wide, for a total of 810 bytes of transported data, including both user
payload and overhead. The first three columns of the frame are the section
and line overhead, known collectively as the *transport overhead*. The bulk
of the frame itself, to the left, is the *synchronous payload envelope (SPE)*,
which is the container area for the user data that is being transported. The
data, previously identified as the payload, begins somewhere in the payload
envelope. The actual starting point will vary, as we will see later. The path
overhead begins when the payload begins; because it is unique to the pay-

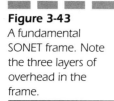

Figure 3-43
A fundamental SONET frame. Note the three layers of overhead in the frame.

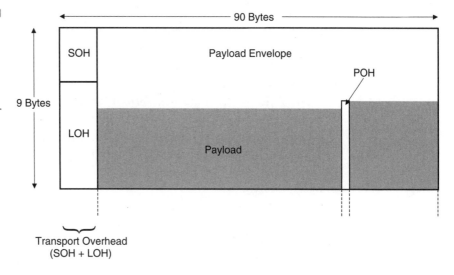

load itself, it travels closely with the payload. The first byte of the payload is in fact the first byte of the path overhead.

A word about nomenclature: Two distinct terms are used, often (incorrectly) interchangeably. The terms are *Synchronous Transport Signal* (STS) and *Optical Carrier Level* (OC). They are used interchangeably because although an STS-1 and an OC-1 are both 51.84 Mbps signals, one is an electrically framed signal (STS) while the other describes an optical signal (OC). Keep in mind that the signals that SONET transports usually originate at an electrical source such as a T1. These data must be collected and multiplexed at an electrical level before being handed over to the optical transport system. The optical networking part of the SONET system speaks in terms of OC.

The SONET frame is transmitted serially on a row-by-row basis. The SONET multiplexer transmits (and therefore receives) the first byte of row one, all the way to the ninetieth byte of row one. It then wraps to transmit the first byte of row two, all the way to the ninetieth byte of row two, and so on, until all 810 bytes have been transmitted.

Because the rows are transmitted serially, the many overhead bytes do not all appear at the beginning of the transmission of the frame; instead, they are peppered along the bitstream like highway markers. For example, the first two bytes of overhead in the section overhead are the framing bytes, followed by the single-byte signal identifier. The next 87 bytes are user payload, followed by the next byte of section overhead; in other words,

87 bytes of user data take place between the first three section overhead bytes and the next one! The designers of SONET were clearly thinking the day they came up with this, because each byte of data appears just when it is needed.

Because of the unique way that the user's data is mapped into the SONET frame, the data can actually start pretty much anywhere in the payload envelope. The payload is always the same number of bytes, which means that if it starts late in the payload envelope, it may well run into the payload envelope of the next frame! In fact, this happens more often than not, but it's okay. SONET is equipped to handle this odd behavior. We'll discuss it shortly.

SONET Bandwidth

The SONET frame consists of 810 eight-bit bytes and, like the T1 frame, is transmitted once every 125 μsec (8,000 frames per second). Doing the math, this works out to an overall bit rate of

$$810 \text{ bytes/frame} \times 8 \text{ bits/byte} \times 8,000 \text{ frames/second} = 51.84 \text{ Mbps}$$

This is the fundamental transmission rate of the SONET STS-1 frame.

That's a lot of bandwidth. 51.84 Mbps is slightly more than a 44.736 Mbps DS-3, a respectable carrier level by anyone's standard. What if more bandwidth is required, however? What if the user wants to transmit multiple DS-3s, or perhaps a single signal that requires more than 51.84 Mbps, such as a 100 Mbps Fast Ethernet signal? Or, for that matter, a payload that requires *less* than 51.84 Mbps! In those cases, we have to invoke more of SONET's magic.

The STS-N Frame

In situations where multiple STS-1s are required to transport multiple payloads, all of which fit in an STS-1's payload capacity, SONET enables the creation of what are called *STS-N frames*, where N represents the number of STS-1 frames that are multiplexed together to create the frame. If three STS-1s are combined, the result is an STS-3. In this case, the three STS-1s are brought into the multiplexer and *byte interleaved* to create the STS-3, as shown in Figure 3-44. In other words, the multiplexer selects the *first*

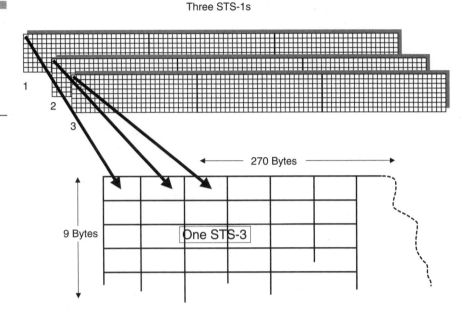

Figure 3-44
Byte interleaving in
SONET. Each lower
bit rate frame is
interleaved with the
others to create a
super-rate frame.

Three STS-1s

1

2

3

270 Bytes

9 Bytes

One STS-3

byte of frame one, followed by the *first* byte of frame two, followed by the
first byte of frame three. Then it selects the *second* byte of frame one, followed by the *second* byte of frame two, followed by the *second* byte of
frame three, and so on, until it has built an interleaved frame that is now
three times the size of an STS-1: 9×270 bytes instead of 9×90. Interestingly (and impressively!), the STS-3 is still generated 8,000 times per
second.

The technique described previously is called a *single-stage multiplexing
process,* because the incoming payload components are combined in a single
step. Also, a two-stage technique is commonly used. For example, an STS-12
can be created in either of two ways. Twelve STS-1s can be combined in a
single-stage process to create the byte interleaved STS-12. Alternatively,
four groups of three STS-1s can be combined to form four STS-3s, which can
then be further combined in a second stage to create a single STS-12. Obviously, two-stage multiplexing is more complex than its single-stage cousin,
but both are used.

Special note: The overall bit rate of the STS-N system is $N \times$ STS-1.
However, the *maximum* bandwidth that can be transported is STS-1, but N
of them can be transported. This is analogous to a channelized T1.

The STS-Nc Frame

Let's go back to our Fast Ethernet example mentioned earlier. In this case, 51.84 Mbps is inadequate for our purposes because we have to transport the 100 Mbps Ethernet signal. For this, we need what is known as a *concatenated signal*. One thing you can say about SONET: It doesn't hurt for polysyllabic vocabulary.

The word concatenate means "to string together," which is exactly what we do when we need to create what is known as a *super-rate frame*—in other words, a frame capable of transporting a payload that requires more bandwidth than an STS-1 can provide, such as our 100 Mbps Fast Ethernet frame. In the same way that an STS-N is analogous to a channelized T1, an STS-Nc is analogous to an *unchannelized* T1. In both cases, the customer is given the full bandwidth that the pipe provides; the difference lies in how the bandwidth is parceled out to the user.

Overhead Modifications in STS-Nc Frames

When we transport multiple STS-1s in an STS-N frame, we assume that they may arrive from different sources. As a result, each frame is inserted into the STS-N frame with its own unique set of overhead. When we create a concatenated frame, though, the data that will occupy the combined bandwidth of the frame is derived from the same source; if we pack a 100 Mbps Fast Ethernet signal into a 155.53 Mbps STS-3c frame, there's only one signal to pack. It's pretty obvious, then, that we don't need three sets of overhead to guide a single frame through the maze of the network.

For example, each frame has a set of bytes that keeps track of the payload within the synchronous payload envelope. Since we only have one payload, we can eliminate two of them, and the path overhead that is unique to the payload can similarly be reduced, since there is a column of it for each of the three formerly individual frames. In the case of the pointer that tracks the floating payload, the first pointer continues to perform that function; the others are changed to a fixed binary value that is known to receiving devices as a *concatenation indication*. The details of these bytes will be covered later in the overhead section.

Transporting Subrate Payloads: Virtual Tributaries

When a SONET frame is modified for the transport of subrate payloads, it is said to carry *virtual tributaries*. Simply put, the payload envelope is chopped into smaller pieces that can then be individually used for the transport of multiple lower-bandwidth signals.

Creating Virtual Tributaries

To create a "virtual tributary-ready" STS, the synchronous payload envelope is subdivided. An STS-1 comprises 90 columns of bytes, four of which are reserved for overhead functions (section, line, and path). That leaves 86 for the actual user payload. To create virtual tributaries, the payload capacity of the SPE is divided into seven 12-column pieces, called *virtual tributary groups*. Math majors will be quick to point out that $7 \times 12 = 84$, leaving 2 unassigned columns. These columns, shown in Figure 3-45, are indeed unassigned and are given the rather silly name of *fixed stuff*.

Figure 3-45
SONET's fixed stuff. You'd think with all the creativity that suffuses Bell Labs they could have come up with a better name.

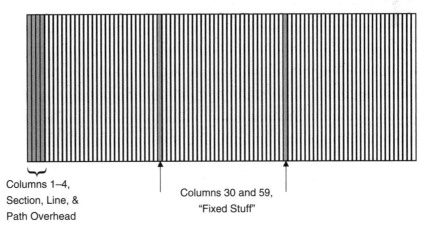

Columns 1–4,
Section, Line, &
Path Overhead

Columns 30 and 59,
"Fixed Stuff"

Now comes the fun part. Each of the virtual tributary groups can be further subdivided into one of four different VTs to carry a variety of payload types, as shown in Figure 3-46. A VT1.5, for example, can easily transport a 1.544 Mbps signal within its 1.728 Mbps capacity, with a little room left over. A VT2, meanwhile, has enough capacity in its 2.304 Mbps structure to carry a 2.048 Mbps European E1 signal, with a little room left over. A VT3 can transport a DS-1C signal, while a VT6 can easily accommodate a DS-2, again, each with a little room left over.

One aspect of virtual tributaries that must be mentioned is the mix-and-match nature of the payload. Within a single SPE, the seven VT groups can carry a variety of different VTs. However, each VT group can carry only one VT type.

That "little room left over" comment is, by the way, one of the key points that SONET and SDH detractors point to when criticizing them as legacy technologies. Such critics claim that in these times of growing competition and the universal drive for efficiency SONET and SDH are inordinately wasteful of bandwidth, given that they were designed when the companies that delivered them were monopolies and less concerned about such things than they are now. We will discuss this issue in a later section of the book. For now, though, suffice it to say that one of the elegant aspects of SONET is its capability to accept essentially any form of data signal, map it into standardized positions within the SPE frame, and transport it efficiently

Figure 3-46
SONET virtual
tributary payload
transport

VT Type	Columns/VT	Bytes/VT	VTs/Group	VTs/SPE	VT Bandwidth
VT1.5	3	27	4	28	1.728
VT2	4	36	3	21	2.304
VT3	6	54	2	14	3.456
VT6	12	108	1	7	6.912

and at very high speed to a receiving device on the other side of town or the other side of the world.

Creating the Virtual Tributary Superframe

You will recall that when DS-1 frames are transmitted through modern networks today, they are typically formatted into extended superframes in order to eke additional capability out of the comparatively large percentage of overhead space that is available. When DS-1 or other signals are transported via an STS-1 formatted into virtual tributary groups, four consecutive STS-1s are grouped together to create a single VT superframe, as shown in Figure 3-47. To identify the fact that the frames are behaving as a VT superframe, certain overhead bytes are modified to indicate the change.

SONET Summary

Clearly, SONET is a complex and highly capable standard designed to provide high-bandwidth transport for legacy and new protocol types alike. The

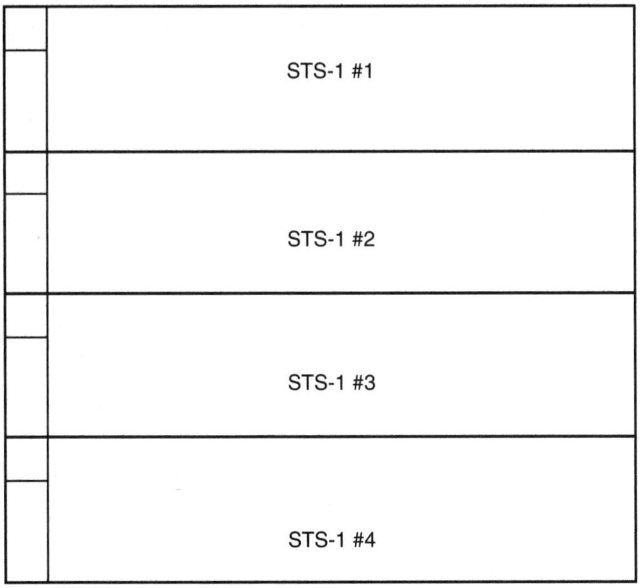

Figure 3-47
A VT superframe

STS-1 #1

STS-1 #2

STS-1 #3

STS-1 #4

overhead that it provisions has the capability to deliver a remarkable collection of network management, monitoring, and transport granularity.

The European SDH shares many of SONET's characteristics, as we will now see. SONET, you will recall, is a limited North American standard for the most part. SDH, on the other hand, provides high-bandwidth transport for the rest of the world.

Most books on SONET and SDH cite a common list of reasons for their proliferation. These reasons include a recognition of the importance of the global marketplace and a desire on the parts of manufacturers to provide devices that will operate in both SONET and SDH environments. Such reasons also include the global expansion of ring architectures, a greater focus on network management and the value that it brings to the table, and a massive, unstoppable demand for more bandwidth. One must also take into account an increasing demand for high-speed routing capability to work hand in glove with transport; the deployment of DS-1, DS-3, and E1 interfaces directly to the enterprise customer as access solutions; a growth in demand for broadband access technologies such as cable modems, the many flavors of DSL, and two-way satellite connectivity; and the ongoing replacement of traditional circuit-switched network fabrics with packet-based transport and mesh architectures. A renewed focus on the SONET and SDH overhead with an eye toward using it more effectively and the convergence of multiple applications on a single, capable, high-speed network fabric are also contributing factors. Most visible among these is the hunger for bandwidth. According to consultancy RHK, global volume demand will grow from approximately 350,000 terabytes of transported data per month in April 2000 to more than 16 million terabytes of traffic per month in 2003. And who can argue?

SDH Nomenclature

Before launching into a functional description of SDH, it would be good to first cover the differences in naming conventions between the two. This will help to dispel confusion (hopefully!).

The fundamental SONET unit of transport uses a 9-row by 90-column frame that comprises 3 columns of section and line overhead, 1 column of path overhead, and 87 columns of payload. The payload, which is primarily user data, is carried in a payload envelope that can be formatted in various ways to make it carry a variety of payload types. For example, multiple SONET STS-1 frames can be combined to create higher-rate systems for transporting multiple STS-1 streams, or a single higher-rate stream created from the combined bandwidth of the various multiplexed components. Conversely, SONET can transport subrate payloads, or virtual tributaries, which operate at rates slower than the fundamental STS-1 SONET rate. When this is done, the payload envelope is divided into virtual tributary groups, which can in turn transport a variety of virtual tributary types.

In the SDH world, similar words apply, but they are different enough that they should be discussed. As you will see, SDH uses a fundamental transport container that is three times the size of its SONET counterpart. It is a 9-row by 270-column frame that can be configured into 1 of 5 container types, typically written C-n (where C means container). N can be 11, 12, 2, 3, or 4; they are designed to transport a variety of payload types.

When an STM-1 is formatted for the transport of virtual tributaries (known as virtual containers in the SDH world), the payload pointers must be modified. In the case of a payload that is carrying virtual containers, the pointer is known as an *Administrative Unit type 3* (AU-3). If the payload is *not* structured to carry virtual containers but is instead intended for the transport of higher-rate payloads, then the pointer is known as an *Administrative Unit type 4* (AU-4). Generally speaking, an AU-3 is typically used for the transport of North American Digital Hierarchy payloads and AU-4 is used for European signal types.

The SDH Frame

To understand the SDH frame structure, it is first helpful to understand the relationship between SDH and SONET. Functionally, they are identical: In both cases, the intent of the technology is to provide a globally standardized transmission system for high-speed data. SONET is indeed optimized for the T1-heavy North American market, while SDH is more applicable to Europe. Beyond that, however, the overhead and design considerations of the two are virtually identical, but some key differences exist.

Perhaps the greatest difference between the two lies in the physical nature of the frame. A SONET STS-1 frame comprises 810 total bytes for an overall aggregate bit rate of 51.84 Mbps, perfectly adequate for the North American 44.736 Mbps DS-3. An SDH STM-1 frame, however, designed to transport a 139.264 Mbps European E-4 or CEPT-4 signal must be larger if it is to accommodate that much bandwidth; it clearly won't fit in the limited space available in an STS-1. An STM-1, then, operates at a fundamental rate of 155.52 Mbps, enough for the bandwidth requirements of the E-4. This should be where the déjà vu starts to kick in. Perceptive readers will remember that the 155.52 Mbps number from our discussions of the SONET STS-3, which offers *exactly* the same bandwidth. An STM-1 frame

Figure 3-48
An SDH STM-1 frame

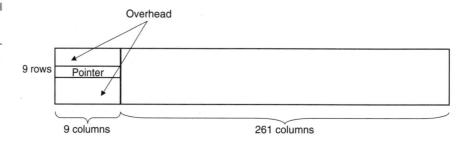

is shown in Figure 3-48. It is a byte interleaved, 9-row by 270-column frame, with the first 9 columns devoted to overhead and the remaining 261 devoted to payload transport.

A comparison of the bandwidth between SONET and SDH systems is also interesting. The fundamental SDH signal is exactly *three times* the bandwidth of the fundamental SONET signal, and this relationship continues all the way up the hierarchy.

STM Frame Overhead

The overhead in an STM frame is similar to that of an STS-1 frame, although the nomenclature varies somewhat. Instead of section, line, and path overhead to designate the different regions of the network that the overhead components address, SDH uses *regenerator section overhead* (RSOH), *multiplex section overhead* (MSOH), and path overhead, as shown in Figure 3-49. The RSOH occupies the first three rows of nine bytes, and the MSOH, the final five. Row four is reserved for the pointer. As in SONET, the path overhead floats gently on the payload tides, rising and falling in response to phase shifts. Functionally, these overhead components are identical to their SONET counterparts.

Overhead Details

Because an STM-1 is three times as large as an STS-1, it has three times the overhead capacity—nine columns instead of three (plus path overhead). The first row of the RSOH is its SONET counterpart, with the exception of the last two bytes, which are labeled as being reserved for national use and are specific to the *Post, Telephone, and Telegraph* (PTT) administration that implements the network. In SONET, they are not yet assigned. The second

Figure 3-49
SDH overhead

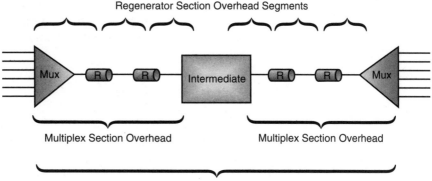

Regenerator Section Overhead Segments

Multiplex Section Overhead   Multiplex Section Overhead

Path Overhead

row is different from SONET in that it has three bytes reserved for media-dependent implementation (differences in the actual transmission medium, whether copper, coaxial, or fiber) and the final two are reserved for national use. As before, they are not yet definitively assigned in the SONET realm.

The final row of the RSOH also sports two bytes reserved for media-dependent information while they are reserved in SONET. All other regenerator section/section overhead functions are identical between the two.

The MSOH in the SDH frame is almost exactly the same as that of the SONET line overhead, with one exception. Row nine of the SDH frame has two bytes reserved for national administration use. They are reserved in the SONET world.

The pointer in an SDH frame is conceptually identical to that of a SONET pointer, although it has some minor differences in nomenclature. In SDH, the pointer is referred to as an *Administrative Unit* (AU) pointer, referring to the standard naming convention described earlier.

Metro Interworking: SONET and Ethernet

Ethernet has been around for a long time and, as stated earlier, is understood, easy to use, and trusted. SONET, too, has been around for quite some time and is equally well accepted and understood. There has been talk of late about Ethernet replacing SONET as the next major transport protocol for the metro arena. Remember, though, that a difference exists between the two: SONET is a physical-layer standard (layer one) and Ethernet is a data-link-layer standard (layer two). Although Ethernet can serve as a framing standard for the transport of user content, it lacks a number of the features of SONET such as QoS control and embedded management tools.

Resilient Packet Ring (RPR)

RPR offers an alternative for the transport of data traffic over a shared network infrastructure with fault detection and recovery within 50 milliseconds. When deployed in the metro environment, users benefit from more efficient, dynamic transport that can adapt relatively easily to changing network loads while also being scalable and therefore capable of adapting to an evolving network over time.

The benefits of RPR are now widely known. They include the following:

- Scalability (as many as 256 nodes per ring)
- Resilient fault detection and correction via a dual self-healing ring with a 50-millisecond switchover capability
- Efficient payload transport (because of spatial reuse)
- Lower transport costs (shared ring)
- Higher performance than switched Ethernet

RPR Evolution RPR evolved from work performed separately but roughly simultaneously by a collection of vendors that were all trying to create the ideal high-speed metro transport solution. Today standards have emerged from both the IEEE (802.17) and the IETF (IP over RPR), but the original work performed by Cisco, Nortel, Dynarc, Luminous Networks, and Lantern Communications still provides the overarching guidance behind the standardization effort. As a result, the RPR concept has evolved from a proprietary scheme first posited by Cisco into more of a standardized, widely accepted transport methodology. An industry support group, the RPR Alliance (www.RPR.org), was recently formalized to push the widespread use of the RPR standard throughout the industry as an alternative to traditional transport schemes.

RPR: How It Works A resilient packet ring comprises multiple nodes that share access to a bidirectional ring. Nodes send data across the ring using a specific MAC protocol created for RPR. The goal of the RPR topology is to interconnect multiple-node ring architecture that is media independent for efficiency purposes.

RPR has a number of advantages over SONET/SDH, which require the pre-allocation of transport resources around the entire ring, such as VT1.5, STS-1, or STM-1. RPR, on the other hand, can transmit a packet partially across an arc of the ring, but it does not have to pre-allocate network resources; it only uses what it needs at the time it has data to insert, using what is called a *buffer-insertion ring mechanism*. As a result, network nodes can reuse available ring bandwidth by transmitting data within the same bandwidth space through other nonoverlapping nodes. This is the genesis of the term *spatial reuse*: Two different nodes can simultaneously transmit a burst of data across different segments of the ring, allocating the entire available bandwidth of the ring during the transmission. Because they use different spatial segments of the transport medium, the two transmitted entities do not overlap and therefore do not interfere with one another. This technique dramatically increases transmission efficiency.

Destination Stripping   The Spatial Reuse Protocol requires the implementation of a technique called *destination stripping*. In destination stripping, the destination node removes the ingress packet from the ring so that it doesn't continue to recirculate around the ring, eating up bandwidth and reducing the overall efficiency of the RPR concept. Furthermore, although SONET/SDH may reserve a portion of ring bandwidth for protection (the entire second ring or a portion of the primary ring, depending on the implementation), RPR enables a node to transmit both data and control information in either direction according to its knowledge of the topology of the network and its desire to use the shortest possible path from the source to the destination.

   This technique has a slight downside, however; if the ring should wrap due to a cable cut or other physical layer failure, ring throughput can decrease due to the potential loss of some of the resources of the network. In modern networks, this happens so infrequently that it is not considered a major failure point. Furthermore, newer metro topologies that provision multiple redundant paths via meshed rings further reduce this liability.

The RPR MAC Scheme   In RPR, both pass-through and added (new add-drop) traffic go through a common point at each node in the ring. The RPR MAC protocol determines whether to transmit pass-through traffic or add traffic based on buffer fills and the traffic priority at each node. RPR utilizes a protocol in which data is encapsulated in a MAC-layer header that is processed at each node and used to determine how to handle an arriving packet and what class of service to assign to the packet within a node.

As a MAC-layer protocol, RPR is physical layer independent. Because of this, manufacturers and service providers want to be able to connect RPR nodes via both legacy optical transport schemes (SONET) and/or newer protocols (10 Gbps Ethernet). They believe that RPR should result in a less expensive transport alternative and it is currently believed to be one of the principal long-term drivers behind Ethernet as a transport protocol.

The long-term goal of the RPR strategy is to facilitate the reuse of 10 Gbps PHY, without reliance on the 10 Gbps MAC. This results in one of the minor disadvantages: It is not possible to achieve interoperability between an RPR node and a 10 Gbps Ethernet MAC/PHY node. As a result, IEEE 802.17 competes with 10 Gbps Ethernet MAC/PHY transport.

Where's the Market?   Most vendors today view RPR as an alternative ring-access mechanism. Target markets in the metro space therefore include DSLAM vendors, *Passive Optical Network* (PON) vendors, and other manufacturers that see the benefit of an alternative to SONET/SDH.

Optical Burst Switching (OBS)

A related technology is *optical burst switching* (OBS). It supports real-time multimedia traffic across the Internet, a natural pairing since the Internet itself is bursty. OBS is particularly effective for high-demand applications that require a high bit rate, low latency, and short-duration transmission characteristics.

OBS uses a one-way reservation technique so that a burst of user data, such as a cluster of IP packets, can be sent without having to establish a dedicated path prior to transmission. A control packet is sent first to reserve the wavelength, followed by the traffic burst. As a result, OBS avoids the protracted end-to-end setup delay and also improves the utilization of optical channels for variable bit rate services. By combining the advantages of optical circuit switching and those of optical packet or cell switching, OBS facilitates the establishment of a flexible, efficient, and bandwidth-rich infrastructure.

Mesh Networks

With all the discussion about ring networks, it is important to also discuss mesh networks. Mesh topologies are typically encountered in WANs because they can easily be deployed in that environment. They are *not* typically seen in metro environments. That, however, is changing. Because of the complex route diversity that mesh networks offer, they are beginning to be taken seriously in metro networking. They are extremely flexible and after all are really nothing more than an interlaced collection of point-to-point connections that can be managed as a series of interconnected rings. They offer higher-speed provisioning, much better use of available bandwidth (since *all* users have access to *all* bandwidth, with nothing reserved for failure compensation), and no restrictions on topology or growth. They enable network managers to perform highly granular traffic engineering from any point on the network to enable the high-speed provisioning of services.

The most common mesh medium is optical fiber, although T1, E1, and other *Plesiochronous Digital Hierarchy* (PDH) technologies are also used. Traditional microwave has long been used in cities for point-to-point interconnection, although it is typically a second choice due to its susceptibility to environmental fade, its licensing requirements, and its tendency to quit working when a building sprouts up in the middle of the shot.

Freespace Optics

Another technology that should be watched is *freespace optics*. Freespace optical solutions rely on open-air laser transmissions between buildings to create low-cost metropolitan networks. Most industry estimates today claim that more than 80 percent of all business buildings are not served by fiber. Freespace optics, therefore, would seem to be a good alternative solution and can be used as mesh alternatives.

Several companies have entered the freespace game, including AirFiber with its 622 Mbps OptiMesh™ product and Terabeam, which provides wireless optical-like connectivity at 5, 10, and 100 Mbps. Both companies offer high degrees of survivability and reliability.

Input/Output (I/O) Interfaces

Driven by the burgeoning I/O demands of bandwidth-hungry applications in the metropolitan area environment and the processors that execute them, I/O standards have evolved rapidly in the last couple of years and now offer tremendous capacity along with ancillary functionality that transforms the "dumb interface" into a relatively intelligent entity. The interfaces covered here include Fibre Channel, the various forms of the

Small Computer Systems Interface (SCSI), *High Performance Parallel Inter-*
face (HPPI), *High-Speed Serial Interface* (HSSI), *Peripheral Component*
Interconnect (PCI), *Universal Serial Bus* (USB), and IEEE 1394.

Fibre Channel

Fibre Channel defines a set of standards for a serial I/O bus that supports
a range of port speeds, including 133 Mbps, 266 Mbps, 530 Mbps, 1 Gbps,
and, soon, 4 Gbps. The standard supports point-to-point connections,
switched topologies, and arbitrated loop architecture.

Arbitrated loop, sometimes called FC-AL, is designed to support SAN
applications, specifically connectivity between a host and a storage array
over connections of up to 10 kilometers at speeds as high as 100 Mbps. It
operates in full-duplex, point-to-point, and serial mode. Fibre Channel supports the logical connection of as many as 126 hosts to a storage array, a
tremendous advantage for the SAN industry.

Fibre Channel was developed as the result of the industry-wide recognition of the need for a protocol that would support the low-cost, high-
bandwidth interface requirements of evolving SAN applications. Unlike
SCSI, which was developed by a single vendor and later submitted for standardization, Fibre Channel was a totally cooperative effort that enjoys widespread acceptance. In addition to SANs, Fibre Channel applications include
video, high-speed network transport, avionics, and IP backbone transport.

In addition to high-speed transport and insensitivity to distance, Fibre
Channel offers small connectors, extremely efficient bandwidth utilization,
broad availability, support for a range of cost and performance models, support for a variety of interfaces, including IP, SCSI, HIPPI, and various forms
of audio and video protocols.

Small Computer System Interface (SCSI)

SCSI, pronounced *scuzzy*, is a collection of ANSI standards that define I/O
parameters for the connection of storage or other external devices to hosts
through host-to-bus (sometimes called host-bus) adapters. Originally
intended for small computer applications, SCSI has become the most widely
deployed I/O bus in the world, although many believe that Fibre Channel
has already surpassed its lead and USB is closing fast. Whatever the case,
SCSI is designed to replace the traditional *Enhanced Integrated Develop-*
ment Environment (EIDE) interface.

The actual speed of the SCSI interface is elusive because of different connector sizes and the variety of implementations (listed in the following sections). The original SCSI hard drive specifications defined three interfaces: the 50-pin Narrow Ultra SCSI at 20 MBps, the 68-pin Wide Ultra SCSI at 40 MBps, and the 80-pin Wide Ultra SCSI, also at 40 MBps.

SCSI comes in a variety of forms, including Fast SCSI, Serial SCSI, Ultra SCSI, Ultra2 SCSI, Ultra3 SCSI, Wide SCSI, and iSCSI. We will present each of these here.

Fast SCSI   Fast SCSI nominally delivers 10 MB of data per second. A variant, Wide Fast SCSI, supports 16-bit data and transfers 20 MBps. Narrow Fast SCSI operates at 10 MBps.

Serial SCSI   Serial SCSI defines an interface that transmits serial data instead of multiconductor parallel data. Interestingly, IEE 1394 (discussed later) and Fibre Channel (discussed earlier) are examples of serial SCSI implementations.

Ultra SCSI   Ultra SCSI supports data transfer rates of 20 MBps over distances ranging from 1.5 meters to 25 meters, depending on the bus type. Ultra SCSI defines both narrow (8 data bits) and wide (16 data bits) buses. A narrow Ultra SCSI bus transfers data at 20 MBps. The wide Ultra SCSI bus transfers data at a maximum of 40 MBps in both the 68- and 80-pin versions.

Ultra2 SCSI   The Ultra2 SCSI interface has a maximum data transport speed of 40 MBps over distances of up to 25 meters. The standards define both narrow (8 data bits) and wide (16 data bits) buses. The narrow bus transfers data at 40 MBps, while the wide bus reaches 80 MBps.

Ultra3 SCSI   Like the wide version of Ultra2 SCSI, Ultra3 can transfer data at 80 MBps over distances up to 12 meters. The Ultra3 specifications define a wide, 16-bit bus that can achieve 160 MBps data transfer rates.

iSCSI   iSCSI is an acronym for Internet SCSI, a new storage networking standard for linking data storage facilities over an IP network. Because of the universal deployment of IP networks, iSCSI can transmit data over LANs, WANs, or the global Internet and can facilitate the creation of true location-independent data storage and retrieval. The standard is considered to be something of a "next-generation Fibre Channel," given that Fibre Channel has become the standard for deployment in SAN installations.

High Performance Parallel Interface (HPPI)

HPPI (pronounced as you would expect—*hippy*) defines an 800 Mbps I/O interface targeted primarily at supercomputer networks. It enjoyed significant uptake initially but soon fell into disfavor because of its incapability to support the growing data transfer demands of high-speed networking. It was soon augmented by HIPPI-6400, more commonly known as the *Gigabyte System Network* (GSN) standard, which operates at a respectable 6.4 Gbps.

High-Speed Serial Interface (HSSI)

Cisco and T3plus Networking created HSSI jointly in the late 1990s as a high-speed (52 Mbps) data transport solution for WANs. The interface uses the same number of pins as the wide SCSI-2 interface. It gained acceptance early on for high-speed connectivity between ATM switching elements, routers, LANs, and other network components.

Peripheral Component Interconnect (PCI)

PCI is designed to connect interface modules to a computer system. The standard support consists of 32- and 64-bit parallel data transfers. Originally designed by Intel, PCI was designed to be comparable to the legacy *Video Electronics Standards Association* (VESA) local bus. PCI 2.0, however, is no longer considered a local bus and is in fact independent of microprocessor design. The standard is designed to synchronize with the clock speed of the microprocessor and is now installed on most new desktop computers, including both Pentium and Power PC-based machines.

Various forms of PCI exist, including Compact PCI, Mini PCI, Low-Profile PCI, Concurrent PCI, and PCI-X. PCI-X is substantially different from the others; it represents a relatively new bus technology that increases the internal bus speed from 66 MHz to 133 MHz and beyond. The maximum data transfer rate between processor and peripherals using the current PCI design is 532 MBps.

Universal Serial Bus (USB)

USB is a so-called plug-and-play interface between a computer and add-on devices, including audio players, game controllers, keyboards, telephones, video cameras, hard drives, scanners, and printers. Because USB is designed to be a universal standard, new devices can be added. Developed by Compaq, IBM, DEC, Intel, Microsoft, NEC, and Northern Telecom, USB is rapidly becoming the most widely deployed interface standard.

USB supports data transports up to 12 MBps. This high degree of bandwidth enables it to support such devices as MPEG video playback systems, virtual reality data gloves, and video-editing equipment, in addition to the more common devices listed earlier.

Since 1996, Microsoft Windows has included USB drivers or software designed to work with specific types of I/O devices. Because of its growing popularity, most new computers and many peripheral devices come equipped with USB as a standard interface. As many as 127 devices can be daisy-chained on a single USB port.

Several versions of the USB standard exist. USB 1.0 and 1.1 are designed to interconnect a variety of relatively low-speed peripherals to a PC externally. Operating at 12 Mbps, these versions connect keyboards, mice, modems, *personal digital assistants* (PDAs), digital cameras, scanners, printers, and other standard devices.

On the other hand, USB 2.0, officially announced in January 2002 at the Las Vegas Consumer Electronics Show, operates at a healthy 480 Mbps. At that transfer rate, USB will pose a potential threat to SAN standards such as Fibre Channel and the faster versions of SCSI.

IEEE 1394

IEEE 1394, otherwise known as the High-Performance Serial Bus, is a standard for interconnecting devices to a PC. Similar to USB, as many as 63 devices can be attached to a single interface with data transfer speeds up to 400 Mbps. The IEEE 1394 standard defines a serial bus architecture that has taken on a variety of names, including Apple's FireWire and Sony's i.Link.

The standard is quite complete and offers a variety of advantages, including a common serial PC connector, a thin cable instead of the more common thick parallel cable, data transfer rates as high as 400 Mbps, plug-and-play capabilities, and the capability to daisy-chain devices to a single port.

The 1394 standard requires that devices be located no farther than 4.5 meters from the PC's physical port. Up to 63 devices can be connected in a single chain, each with the 4.5-meter maximum to reduce end-to-end attenuation.

Summary

Because metro represents a crossover point between the access and transport domains—it sits between them, after all—many of the technologies used in these traditional regions also appear in and are utilized in metro networks. SONET/SDH, ATM, and Frame Relay all have a role in the evolving and very strong metro marketplace. New solutions such as high-speed Ethernet transport, RPR, and OBS have also appeared on the scene and are fundamentally changing the way networks operate in metropolitan areas.

In Part 4, "Metro Applications," we discuss the applications that empower metro networking.

Metro
Applications

Having discussed the evolution of metro networks and the technologies that underlie their functionality, we now turn our attention to the applications that are most commonly found within the metro environment. Some of them have been in existence for quite some time; others have emerged with the arrival of metro as a standalone region of the network and a support mechanism for the business environment it serves.

The most common applications on the metro scene are fundamentally based upon storage technologies. As competition has increased in all industries (not just telecom), the need to accurately deliver on customer requests for service has become far more critical than ever before. The law of primacy (whoever gets there first wins) has become a crucial component of business success. Of course, "getting there first" only counts if the service provider gets there first with accurate, correct, and appropriate solutions to a customer's business problem. The applications that have emerged most recently and that have garnered the bulk of the marketplace's attention include data mining, information and knowledge management, customer relationship management, enterprise resource planning, and the various storage solutions that make them possible within the metro domain. Each of these will be discussed in the pages that follow.

Corporate Evolution

The modern corporation has evolved rapidly (and dramatically) in the last few years, an evolution driven largely by technology changes, a power shift from the provider to the customer, and a recognition of the enormous value that knowledge about customers has in the typical corporation. Never has the phrase "knowledge is power" been so true, and companies are scrambling to make knowledge and its use a competitive advantage. Another wonderful expression claims "time is money, but speed is profit." Whoever first satisfies the customer's requirements correctly wins the game.

A number of characteristics guide the evolution of the modern corporation in its drive to satisfy the customer and become the preferred vendor of services. These characteristics include the management model that guides the corporation, the actual structure of the overall corporation, the relationship between the corporation and its competitors, the nature of the value chain that guides the behavior of the corporation, the customer's perception of the corporation and its role, and, finally, the corporation's perception of value and what defines it.

Corporations today have their behavioral roots in the Industrial Age, when corporations relied on the mass production of a commodity to reach critical mass in the market. These corporations were largely supplier driven because their ability to create the products they sold was based entirely on the availability of relatively scarce physical resources. Furthermore, these industries tended to be vertical in nature, creating a single homogeneous product that was in widespread demand across a variety of *other* vertical industries. They owned the entire production process, from the raw material to the distribution and marketing effort.

Businesses are characterized by two key characteristics. One is the degree to which they create value to distinguish them from their competitors; the other is the nature of the resources they use to create their products. For the Industrial Age corporation, six key characteristics defined its behavior in the marketplace. First, the management model that guided the corporation was a top-down, hierarchical structure. Authority was concentrated in the upper tier of the management hierarchy, and knowledge required to make informed decisions was highly compartmentalized.

Second, the internal corporate structure was equally hierarchical, with multiple levels of management that expanded in terms of the number of employees as the levels descended.

Third, the relationship between the corporation and its competitors was somewhat antagonistic. Because these companies sold commodities, price was the only differentiator they could rely on. Therefore, they competed aggressively with each other.

Fourth, the value chain was absolutely linear. On one end of the business, raw materials were delivered in trucks. At each progressive step of the manufacturing process, incremental value was added. Finally, at the end of the process, some high-demand commodity popped out.

Fifth, these corporations were all-powerful and in most cases dictated terms to the customer. For the most part, the customer had no leverage in the buyer-seller relationship and was forced to accept whatever the provider offered them, including the terms of the buy-sell arrangement.

Finally, the definition of value in the supplier corporation was based upon the degree to which the corporation could hold sway over its suppliers and outproduce the competition by offering the most product for the least amount of money. The name of the game was volume.

Today the model is quite different, with the principal goals being to improve operational efficiencies wherever possible and to greatly reduce operational costs. Additionally, the six managerial characteristics no longer apply. The top-down, hierarchical, centrally controlled corporation has been replaced by a flatter organization within which decision-making power is distributed at all levels. Instead of bellicose relationships with competitors, many corporations have crafted cooperative relationships that ensure the survival of all players.

The concept of the overall value chain has changed as well. Instead of being linear, the value chain has become nonlinear. In a sense, value chains have become hyperchains, and customers now perceive the corporation as a peer that is involved as a partner in the success of their business. Furthermore, a value chain of information now addresses the development of knowledge necessary to ensure a competitive advantage.

The most important change is how corporations perceive value. Industrial Age corporations measured it according to their ability to produce large quantities of products. Today's corporations, on the other hand, measure value according to how well they use the knowledge they acquire about their customers, suppliers, and competitors to create advantageous positions for themselves and use that position to deliver superior customer service. Teams have become the organizational norm: They are future-focused, responsive to customer requests, and competitively positioned.

The Importance of Corporate Knowledge

In the new marketplace, knowledge-based corporations rise to the top of the competitive heap because they have learned that knowledge about the customer is the most important internal resource they have. It isn't enough to have the best router, the most bandwidth, or the most survivable network. Unless the product is uniquely targeted at each customer's business issues, it's just another of many similar products.

The need to collect, analyze, and respond to knowledge about customers is critical because if executed properly, knowledge management can become the most critical competitive advantage a company has. The process of collecting the data, storing it so that it can be analyzed, and making decisions based on the knowledge it provides falls under a general family of processes called *enterprise resource planning* (ERP). ERP, if executed properly, leads to extremely effective *customer relationship management* (CRM), the cornerstone of the modern relationship between buyers and sellers.

ERP is a corporate planning and internal communications system that affects the deployment of corporate resources. ERP systems are designed to address planning, transaction processing, accounting, finance, logistics, inventory management, sales order fulfillment, supply chain management, human resources operations and tracking, payroll, and CRM. It serves as the umbrella function for a variety of closely integrated functions that are characteristic of the knowledge-based corporation.

The ERP Process

The process by which ERP yields business intelligence is rather interesting. During a typical business day, interactions with customers yield vast amounts of data that are then stored in corporate databases. The data include records of sales, product returns, service order processing, repair data, notes from meetings with customers, competitive intelligence, supplier information, inferential data from meetings, and so on, all of which are the result of normal business activities.

ERP defines an overarching concept under which are found a number of functions, including data warehousing, data mining, knowledge management, and CRM. It defines the process by which corporate data is machined into a finely honed weapon to achieve a competitive advantage.

The Data Process

The data collected during the ERP process is typically archived in a data warehouse of some sort, typically a disk farm behind a large processor in a data center. At this point, the data is exactly that—an unstructured collection of business records stored in a format easily digestible by a computer and that does not yet have a great deal of strategic value. In order to gain value, the data must somehow be manipulated into *information*. Information is defined as a collection of facts or data points that have value to a user.

The raw material that later becomes information is typically accumulated using a technique called *data warehousing*. Data warehouses store data but have very little to do with the process of converting it to information. Data warehouses make the data more available to the applications that will manipulate it, but they do not take part in the process of converting the data into business intelligence. In fact, according to Brio Corporation, corporate information systems and the users who interact with them form an "information supply chain" that includes the raw material (data), the distribution systems themselves (the data warehouse and corporate network), the manufacturers or producers (the *information technology* [IT] organization), and the consumers (end users).

Storage Options

Storage technologies have evolved dramatically in the last decade. The original storage option was called *direct attached storage* (DAS). In DAS arrangements, the storage media (hard drives, CDs, and so on) are either integral to the server (internally mounted) or are directly connected to one of the servers. *Network-attached storage* (NAS), on the other hand, is an architecture in which a server accesses storage media via a LAN connection. The storage media are connected to another server. Finally, *Storage Area Networking* (SAN) architectures rely on a dedicated storage network that provides access to stored content. In a SAN, multiple servers may have access to the same servers.

DAS has been around for a long time and continues to be widely deployed. Its utility is somewhat limited, however, because by its very nature, it is directly connected to a single device. NAS and SAN solutions offer considerably greater utility and have become the accepted choice for shared database access. Contrary to popular belief, NAS and SAN are complimentary technologies. For example, a NAS system can be attached to a SAN, using SAN for its data storage and management.

Although the two are complimentary storage solutions, SAN offers some advantages over NAS. NAS systems operate at the file level, which results in considerably less efficient operations for transaction-based systems. They rely on a shared network infrastructure that, on the one hand, may help to distribute costs more effectively, but it may also result in network congestion. NAS is easy to implement and equally easy to augment and expand. It is also a relatively low-cost solution for evolving corporate storage requirements. Furthermore, the technologies it relies on are common and familiar to corporate IT personnel.

SANs, on the other hand, operate at the data block level, which makes them highly responsive and ideal for transaction-based storage applications. They rely on a dedicated storage network, which will undoubtedly be more costly but will not add to existing LAN traffic. This is a relatively new solution that can easily be augmented as requirements change.

The advantages of a SAN are significant and are partially the reason for their growing success over NAS solutions. These advantages include the following:

- The conservation of LAN bandwidth by moving storage-related traffic off the data network to a dedicated storage transport infrastructure
- Faster backup and restoration capabilities for critical databases
- Centralized storage management and the potential consolidation of servers
- More efficient allocation of storage capacity to servers based upon need as the demand presents itself
- Reduced server downtime due to the periodic addition of storage capacity
- Support for the storage requirements of heterogeneous data environments
- Ubiquitous connectivity
- A high degree of database availability

Storage has become the new application focal point for enterprise spending. In 2000, corporations spent $11 billion on storage applications and supporting technologies; that number is expected to exceed $23 billion by 2003. According to consultancy Dataquest, IT spending on storage is on the rise in 95 percent of all corporations, to the tune of 25 to 50 percent per year. The principal drivers behind this upswing are an increasing need to access customer data, the presence of more critical data on the enterprise network, an acute awareness of the need for disaster recovery capabilities, and an increasing number of critical applications.

The cost of storage deployment has gone up in lockstep with the demand. One reason for this increase is people related. According to IT surveys, a single employee can effectively manage approximately 100 GB of distributed storage. By extension, 10 employees can manage a single *terabyte* (TB) of data, while 150 employees can manage 15 TB. The number changes dramatically when consolidated (SAN) storage is brought into the picture. A single person can now manage roughly one TB, while 15 people can manage 15 TB, a significant multiplier that results in a 5-to-1 cost savings over a distributed storage management model.

All in all, consolidated or centralized storage solutions result in an overall reduction of the total cost of storage as the result of lower equipment costs due to device sharing, reduced human resources requirements, better resource utilization, rapid storage resource allocation, and an overall reduction of downtime.

So what are the applications that storage solutions make possible? We begin with a discussion of data mining.

Data Mining

Data mining is the process used to convert data into information. It is a technique used to identify, examine, and model corporate data as a way to identify customer, vendor, or competitor behavior patterns that can be used to gain a competitive business advantage. Gartner Group defines data mining as "the process of discovering meaningful new correlations, patterns, and trends by sifting through large amounts of data stored in repositories, using pattern recognition technologies, as well as statistical and mathematical techniques." Aaron Zornes of The META Group defines it as "a knowledge discovery process of extracting previously unknown, actionable information from very large databases."

Data mining came about because of the recognized need to manipulate corporate databases in order to extract meaningful information that supports corporate decision-makers. The process produces information that can be converted into knowledge, which is nothing more than a sense of familiarity or understanding that comes from experience or study. Knowledge yields competitive advantage.

Data mining relies on sophisticated analytical techniques such as intelligence filters, neural networks, decision trees, game theory, and behavioral analysis tools that can be used to build a model of the business environment based on data collected from multiple sources. It yields patterns that identify new or not-yet-emerged business opportunities.

Corporations that implement a data mining application do so for various reasons, including

- **Customer and market activity analysis**   By analyzing the buying and selling patterns of a large population of customers, corporations can identify such factors as what they are buying (or returning), why they are buying (or returning), whether or not their purchases are linked to other purchases or market activities, and a host of other factors. One major retailer claims that their data-mining techniques are so indicative of customer behavior that, given certain conditions in the marketplace that they have the ability to predict, they can dramatically change the sales of tennis racquets by lowering the price of tennis balls by as little as a penny.

- **Customer retention and acquisition**   By tracking what customers are doing and identifying the reasons behind their activities, companies can identify the factors that cause customers to stay or go. Domino's, for example, carefully tracks customer purchases, relying on Caller ID information to build a database of pizza preferences, purchase frequency, and the like. Customers enjoy the personalized service they get from the company when they place orders. Similarly, Domino's can identify customers who ordered once but never ordered again, and send them promotional material in an attempt to regain them as a customer. Amazon.com and Dell use similar techniques to increase sales by carefully analyzing customer-buying patterns and then courting customers with carefully crafted advertising schemes.

- **Product cross-selling and upgrading** Many products are complementary, and customers often appreciate being told about upgrades to the product they currently have or are considering buying. Amazon.com is well known for providing this service to their customers. When a patron has made a book purchase online, they often receive an e-mail several days later from Amazon.com, telling them that the company has detected an interesting trend: People who bought the book that the customer recently purchased also bought the following books (list included) in greater than coincidental numbers. They then make it inordinately easy for the customer to "click here" to purchase the other book or books. This is a good example of the benefit of data mining. This so-called shopping cart model has become the standard for online purchases.

- **Theft and fraud detection** Telecommunications providers and credit card companies may use data mining as a way to detect fraud and assess the effectiveness of advertising. For example, American Express and AT&T will routinely call a customer if they detect unusual or extraordinary usage of a credit or calling card, based on the customer's known calling patterns. The same information is often used to match a customer to a custom-calling plan.

A common theme runs through all these subtasks. Data mining identifies customer activity patterns that help companies understand the forces that motivate those customers. The same data helps the company understand the vagaries of their customers' behavior to anticipate resource demand, increase new customer acquisition, and reduce the customer attrition rate. These techniques are becoming more and more mainstream; Gartner Group predicts that the use of data mining for targeted marketing activities will increase enormously over the course of the next 10 years. A number of factors have emerged that have made this possible, including improved access to corporate data through extensive networking and more powerful, capable processors, as well as easy-to-use statistical analysis tools.

A significant number of companies have arisen in the data-mining field, creating software applications that facilitate the collection and analysis of corporate data. One challenge that has arisen, however, is that many data-mining applications suffer from integration difficulties, don't scale well in large (or growing) systems, or rely on a limited number of statistical analysis techniques. Consequently, users spend far too much time manipulating the data, and far too little time analyzing it.

Another factor that often complicates data-mining activities is the corporate perception of what data mining does. Far too often, decision-makers come to believe that data mining alone will yield the knowledge required to make informed corporate decisions, a perception that is incorrect. Data mining does not make decisions; people do, based on the results of data mining. It is the knowledge and experience of the analysts using data-mining techniques that convert its output into usable information.

Data-mining software must rely on multiple analytical techniques that can be combined and automated to accelerate the decision-making process and that deliver the results in a format that makes the most sense to the user of the application.

Knowledge Management

Data mining yields information that is manipulated through various managerial processes to create knowledge. Data mining is an important process, but the winners in the customer service game are those capable of collecting, storing, and manipulating knowledge—which in turn is distilled from information—to create indicators for action. Ideally, those indicators are shared within the company among all personnel who have the ability to use them to bring about change.

The collection of business intelligence is a significant component of knowledge management and yields a variety of benefits, including improvements in personnel and operational efficiencies, enhanced and flexible decision-making capabilities, more effective responses to market movement, and the delivery of innovative products and services. The process also adds certain less tangible advantages, including a greater familiarity with the customer's business processes as well as a better understanding of the service provider on the part of the customer. By combining corporate goals with a solid technology base, business processes can be made significantly more efficient. Finally, the global knowledge that results from knowledge management yields an enhanced understanding of business operations.

Obstacles to Effective Knowledge Management

Knowledge, while difficult to quantify and even more difficult to manage, is a strategic corporate asset. However, because large swaths of it often exist "in the heads" of the people who create it, an infrastructure must be designed within which it can be stored and maintained. Knowledge must also be archived in a way that makes it possible to deliver it to the right people at the right time, always in the most effective format for each user.

The challenge is that knowledge exists in enormous volumes and grows according to Metcalfe's Law. Bob Metcalfe, cofounder of 3Com Corporation and the coinventor of Ethernet, postulated that the value of the information contained in a network of servers increases as a function of the square of the number of servers attached to the network. In other words, if a network grows from 1 server to 8, the value of the information contained within those networked servers increases 64 times over, not 8. Clearly, the knowledge contained in the heads of a corporation's employees multiplies in the same fashion, increasing its value exponentially.

However, this knowledge growth has practical limitations. According to Daniel Tkach of IBM, the maximum size of a company where people actually know each other and who also have a realistic and dependable understanding of collective corporate knowledge is somewhere between 200 and 300 people. As the globalization of modern corporations and the markets they serve continues, how can a corporation reasonably expect to remain aware of the knowledge they possess, when that knowledge is scattered to the four corners of the planet? Studies have indicated that managers glean two-thirds of the information they require from meetings with other personnel; only one-third comes from documents and other "nonhuman" sources. Clearly, a need exists for a knowledge management technique that employees can use to store what they know so that the knowledge can be made available to everyone.

One answer to this question lies in the evolution of the metropolitan network. The reader will recall that one reason behind the urgency of the metro network is the need to put employees closer to customers. By distributing corporate resources (that is, people) in such a way that they are located closer to customers, the corporation can better respond to the needs of those customers. A small office that is dedicated to a cluster of customers can share information about those customers far more effectively internally than it can with the entire corporation. That information that needs to be shared with the corporation at large can be mined by corporate applications as required.

According to a study conducted recently by Laura Empson, a recognized authority on knowledge management, 78 percent of major U.S. corporations indicated their intent at the time of the survey to implement a knowledge management infrastructure. However, because the definition of knowledge management is still somewhat unclear, companies should spend a considerable amount of time thinking about what it means to manage knowledge and what it will take to do so. Empson observes that knowledge includes experience, judgment, intuition, and values, none of which are easily codifiable.

What must corporations do, then, to ensure that they put into place an adequate knowledge management infrastructure? In theory, the process is simple, but the execution is somewhat more complex. First, corporations must clearly define the guidelines that will drive the implementation of the knowledge management infrastructure within the enterprise. Second, they must have top-down support for the effort, with an adequate explanation of the reasons behind the addition of the capability. Third, although perhaps most important, they must create within the enterprise a culture that places value on knowledge and recognizes that networked knowledge is far more valuable than standalone knowledge within one person's head. Too many corporations have crafted philosophies based upon the belief that knowledge is power, which causes individuals to hoard what they know in the mistaken belief that to do so creates a position of power for themselves. The single greatest barrier to the successful implementation of a knowledge management infrastructure is failure of the organization to recognize the value of the new process and accept it.

Fourth, the right technology infrastructure must be created to ensure that the system works efficiently. Corporations often underestimate the degree to which knowledge management systems can consume computer and network resources. The result is an overtaxed network that delivers marginal service at best. Today a significant number of ERP and knowledge management initiatives fail because corporate IT professionals fail to take into consideration the impact that traffic from these software systems will have on the overall corporate network. The traffic generated from the analysis of customer interactions can be enormous, and if the internal systems are incapable of keeping up, the result will be a failed CRM effort and degradation of other applications as well.

As corporations get larger, it becomes more and more difficult to manage the processes with which they manage their day-to-day operations. These processes include buying, invoicing, inventory control and management, and any number of other functions. Consulting firm AMR estimates that "maverick buying"—that is, the process of buying resources in a piecemeal fashion by individuals instead of through a central purchasing agency—

accounts for as much as a third of their total operating resource expenditures. This also adds a 15 to 27 percent overlay premium on those purchases due to the loss of critical mass from one-off purchasing.

According to studies conducted by Ariba, a successful provider of knowledge and supply-chain management software, corporations often spend as much as a third of their revenues on operating resources: office supplies, IT equipment and services, computers, and maintenance. Each of these resources has associated with it a cost of acquisition that can amount to as much as $150 per transaction when conducted with traditional nonmechanized processes. Systems like those from companies like Ariba can help corporations manage costs by eliminating the manual components in favor of computer and network-driven processes.

Managing Quality of Service (QoS)

Perhaps nothing is more elusive than QoS in networks today. Quality, by definition, is entirely subjective, elusive to codify, and virtually impossible to manage and measure. Many service providers attempt to offer it as a differentiator, defining it seven ways from Sunday when in fact it is the *customer* who should provide the appropriate definition. QoS is a defining component of the relationship between a customer and a supplier; the customer should define the nature of quality while the supplier should strive to meet it.

Several corporations offer products that are specifically targeted at the relationships between telecommunications providers and customers. Many, for example, offer software and hardware combinations that allow network managers to monitor a service provider's circuits to determine whether the provider is meeting *service level agreements* (SLAs). Given the growing interest in SLAs in today's telecommunications marketplace, this application and others like it will become extremely important, as SLAs become significant competitive advantages for service providers.

An Aside: The SLA

SLAs have become the defining performance standard in the network and systems game. Their role is to define and quantify the responsibilities of both the service provider that is being measured and the customer that is purchasing the service provider's solution. Measures detailed include the rights a customer has, the penalties that can be assessed against the service provider for failure to perform according to the SLA contract, and, of course, the nature of the delivered service that must be maintained. The goal of SLAs is to ensure performance, uptime, and maintenance of QoS in environments where networks and systems are critical to the corporation's ability to deliver services to its customers.

The SLA game has become big business. Consultancy IDC has predicted that the overall market for hosted and managed SLAs will grow to more than $849 million by 2004.

Industry analysts have defined four primary areas that SLAs cover in the modern service-driven corporation. These include the actual network, the hosting services that might be delivered by the service provider, the applications themselves that the client corporation is dependent upon, and the customer service function, which includes technical support, software update activities, and help desk functions. Each of these four management areas is characterized by a collection of predefined characteristics of various managed entities. These include

- *Network service*, such as a clear and acceptable definition of network infrastructure, acceptable levels of data loss and latency, a clearly defined level of security, and a list of the services, applications, and data that are protected

- *Hosted services*, such as server availability, server administration and management functions, and database backup and archive schedules

- *Applications*, including application-specific measurements such as the number of downloads, web page hit counts, user transaction counts, and the average response time of user requests for information

- *Customer service issues*, such as the degree to which service-monitoring functions are proactive, the maximum wait time for calls to the help desk, the availability of technical support, and the average wait time for assistance

Supply-Chain Issues

Many of the concerns addressed by the overall ERP process are designed to deal with supply-chain issues. Although supply chains are often described as the relationship between a raw materials provider, a manufacturer, a wholesaler, a retailer, and a customer, it can actually be viewed more generally than that. The supply-chain process begins the moment a customer places an order, often online. That simple event may kick off a manufacturing order, reserve the necessary raw materials and manufacturing capacity, and create expected delivery reports.

A number of companies focus their efforts on the overall supply chain. Most offer an array of capabilities including customer service and support, relationship-building modules, system personalization, brand-building techniques, account management, financial forecasting, product portfolio planning, development scheduling, and product transition planning.

ERP in the Telecomm Space

ERP's promise of greatly improved operational efficiencies didn't initially attract much attention from telecom carriers, particularly the *incumbent local exchange carriers* (ILECs). As regulated carriers with guaranteed rates of return and huge percentages of marketshare, they did not rise quickly to the ERP challenge. Recently, however, their collective interest level has grown, as the capabilities embedded in the ERP suite have become better known.

One factor that has accelerated telecom's interest in ERP is the ongoing stream of mergers and acquisitions within the telecom arena. When mergers occur, a great deal of attention is always focused on their strategic implications, but once the merger has taken place and the excitement has settled down, a great deal of work is still left to do internally—hence the value of such professional services companies as Deloitte & Touche. Not only does the merged company now have multiple networks that must be combined, but they also have multiple support systems that must be integrated to ensure seamless billing, operations, administration, provisioning, repair, maintenance, and all the other functions necessary to properly operate a large network.

Telcos are inordinately dependent upon their *operations support systems* (OSSs) and are therefore more than a little sensitive to the issues they face with regard to backroom system integration. One of the reasons they have been somewhat slow to embrace ERP is the fact that in many cases they have waited for ERP vendors to establish relationships with the telcos' OSS providers to ensure that together they address interoperability concerns before proceeding.

Another concern that has slowed ERP implementation among telecom providers is fear of the unknown. ERP is a nontrivial undertaking, and given the tremendous focus that exists today on customer service, service providers are loathe to undertake any activity that could negatively affect their ability to provide the best service possible. For example, ERP applications tend to require significant computing and network resources if they are to run properly. Telcos are therefore reluctant to deploy them for fear that they could have a deleterious impact on service.

Furthermore, ERP implementation is far from being a cookie-cutter exercise. The process is different every time, and "surprises" are legion. The applications behave differently in different system and application environments, and as a consequence they are somewhat unpredictable. Add to that the fact that even with the best offline testing, a certain amount of mystery always remains with regard to how the system will behave when placed under an online load.

Because of the competitive nature of the marketplace in which telecom providers operate today, and because of the fact that the customer, not the technology, runs the market, service providers are faced with the realization that the customer holds the cards. Whereas there was once a time when the customer might call the service provider to order a single technology-based product, today they are far more likely to come to the service provider with the expectation that they be exactly that—the provider of a full range of services, customized for the customer's particular needs and delivered in the manner most convenient to the customer at that point in time.

Because of their access and transport legacy, incumbent service providers understand technology and what it takes to deploy it. As the market has become more competitive, however, they have had to change the way they view their role in the market. They can no longer rely on technology alone to satisfy the limited needs of their customers, because those same customers have realized that technology is not an end itself; it is an important part of a service package. No longer can the service provider roll out a new technology and expect the market to scurry away and find something to do with it. Their focus must change from being technology-centric to service-centric. This is not bad news; the technology that they are so good at providing simply becomes a component of their all-inclusive services bundle. And with the proliferation of new operational environments such as metro in response to customer demands for service, the need to focus on service as opposed to technological capabilities is of paramount importance for service providers if they are to maintain their competitive advantage with existing customers.

Customer Relationship Management (CRM)

The final component in the ERP family is CRM. CRM is the culmination of the ERP process, a collection of information that results from an analysis of business intelligence, which in turn derives from the data-mining effort. Many companies offer CRM software today, including Peoplesoft, IBM, Siebel, Hewlett-Packard, Oracle, and others. Their goal is to cement the relationship between the supplier and the customer, creating a cooperative environment between the two.

Putting It All Together

If we consider the entire ERP process, a logical flow emerges, shown in Figure 4-1, which begins with normal business interactions with the customer.

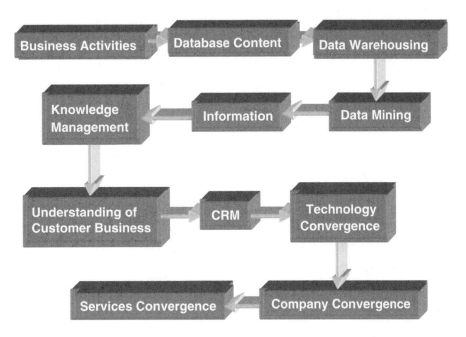

Figure 4-1
The flow of data in the ERP process. Data becomes information, which in turn becomes knowledge. This overall process is used for effective CRM.

As the customer makes purchases, queries the company for information, buys additional add-on features, and requests repair or maintenance services, database entries accumulate that result in vast stores of uncorrelated data. The data is housed in data warehouses, often accessible through SAN, interconnected via high-speed trunking architectures such as Hyper-Channel.

From the data warehouses, corporate data-mining applications retrieve the data and massage it logically to identify trends and indicative activities that help the company understand the drivers behind its customers' behaviors. Thus, they can be anticipated and acted upon before the fact. The data-mining process, then, yields information that now has enhanced value.

The information derived from data mining is manipulated in a variety of ways and combined with other information to create knowledge. Knowledge is created whenever information is mixed with human experience to further enhance its value. The knowledge can then be managed to yield an even more accurate understanding of the customer.

Once customers' behaviors are understood, strategies can be developed that will help the service provider anticipate what each customer will require in the future. This leads to an enhanced relationship between the company and its customers, because the company is no longer in the position of simply responding to customer requests, but in fact it can predict what the customer will require. This represents CRM at its finest.

Consider this model, then, within the context of the telecommunications service provider space. Once the service provider has developed an enhanced understanding of and relationship with the customer, and they know what the customer base will be looking for in the near and long-term in the way of products and services, the service provider can develop a technology plan that will help to ensure that they have the right technologies in the right places at the right time to satisfy those service requests when they arrive.

Given the accelerated pace of the telecommunications market, however, service providers do not have the time to develop new technologies in house. Instead, they do the next best thing, which is to go to the market, identify another company that offers the technology they require, and either buy the company or form an exclusive, strategic alliance with them. The service provider gains the technological capabilities they require to satisfy their customer, the technology provider gains an instantaneous and substantial piece of marketshare, and the customer's requirements are not only met, but they are anticipated.

Network Management in Metro Networks

Because of the diverse audiences that require network performance information and the importance of SLAs, the data collected by network management systems must be manipulatable so that it can be formatted for different sets of corporate eyes. For the purposes of monitoring performance relative to SLAs, customers require information that details the performance of the network relative to the requirements of their applications. For network operations personnel, reports must be generated that detail network performance to ensure that the network is meeting the requirements of the SLAs that exist between the service provider and the customer. Finally, for the needs of sales and marketing organizations, reports must be available that enable them to properly represent the company's abilities to customers and to anticipate requirements for network augmentation and growth.

For the longest time, the *Telecommunications Management Network* (TMN) has been considered the ideal model for network management. As the network profile has changed, however, with the steady migration to IP and a renewed focus on service rather than technology, the standard TMN philosophy has begun to appear somewhat tarnished.

Originally designed by the *International Telecommunication Union— Telecommunications Standardization Sector* (ITU-T), TMN is built around the *Open Systems Interconnection* (OSI) Model and its attendant standards, which include the *Common Management Information Protocol* (CMIP) and the *Guidelines for the Development of Managed Objects* (GDMO).

TMN employs a model, shown in Figure 4-2, comprising a network element layer, an element management layer, a network management layer, a service management layer, and a business management layer. Each has a specific set of responsibilities closely related to those of the layers that surround it.

The network element layer defines each manageable element in the network on a device-by-device basis. Thus, the manageable characteristics of each device in the network are defined at this functional layer.

The element management layer manages the characteristics of the elements defined by the network element layer. The information found here includes activity log data for each element. This layer houses the actual element management systems responsible for the management of each device or set of devices on the network.

Figure 4-2
The TMN layered
model

The network management layer has the capability to monitor the entire network based upon information provided by the element management layer. The services management layer responds to information provided by the network management layer to deliver such service functions as accounting, provisioning, fault management, configuration, and security services.

Finally, the business management layer manages the applications and processes that provide strategic business planning and the tracking of customer interaction vehicles such as SLAs.

OSI, while highly capable, has long been considered less efficient than *Internet Engineering Task Force* (IETF) management standards, and in 1991, market forces began to effect a shift. That year, the Object Management Group was founded by a number of computer companies, including Sun, Hewlett-Packard, and 3Com, and together they introduced the *Common Object Request Broker Architecture* (CORBA). CORBA is designed to be vendor independent and built around object-oriented applications. It enables disparate applications to communicate with each other, regardless of physical location or vendor. Although CORBA did not achieve immediate success, it has now been widely accepted, resulting in CORBA-based development efforts among network management system vendors. This may seem to fly in the face of the OSI-centric TMN architecture, but it really doesn't. TMN is more of a philosophical approach to network management and does not specify technological implementation requirements. Thus, a conversion from CMIP to the *Simple Network Management Protocol* (SNMP), or the implementation of CORBA, does not affect the overall goal of TMN.

Today CORBA is the accepted standard for management interfaces and is widely deployed throughout telecommunications networks. Large numbers of vendors offer CORBA-compliant management systems, and advances in interoperability occur almost weekly. In the metro environment, a number of vendors, discussed in Part 5, "Players in the Metro Game," have crafted innovative management schemes for the increasingly common mesh networks found there.

Summary

The truth is, applications found in the metro environment are no different than those found in traditional access and transport regions of the network. However, because the metro domain serves as a transition zone between access and transport, the applications tend to be somewhat specialized in that they serve both regions. QoS in metro networks tends to be a challenge because in many cases a metro network is used to span the connectivity gap between work groups in the same company that are now separated by distance, when before they shared connectivity via a common LAN. The metro network must therefore achieve QoS levels that emulate the connectivity performance that users enjoy when attached to the same LAN.

Obviously, management plays a key role here. IT groups within corporations must recognize the unique characteristics of metropolitan networks and the customers they serve while at the same time recognizing that they must seamlessly interconnect legacy (*Synchronous Optical Network / Synchronous Digital Hierarchy* [SONET/SDH] and T1) and emerging (Ethernet) technologies within the same overall network. This represents a significant challenge for companies wanting to play in the metro space. In Part 5, we meet some of them.

5

Players in the Metro Game

The proliferation of demand for metro services is going to result in a significant increase in metro spending, a fact that will affect all three levels of the metro networking hierarchy: the component manufacturers, the systems manufacturers, and the service providers (see Figure 5-1). Between now and 2006, metro spending in North America will grow from approximately $400 million to $3 billion, with roughly $2.5 billion of that in products and half a billion in services.

Primary reasons for this increase include performance improvement, the demand for increased network availability, and network robustness and redundancy. Interestingly, the projected spending trends reflect an evolution away from traditional connectivity (T1 and DSL) in favor of Ethernet. Because most corporate traffic begins and ends as Ethernet frames today, a strong push exists to extend its use to the *wide area network* (WAN), hence the proliferation of Ethernet service providers. Furthermore, the evolving applications described earlier—*storage area networks* (SANs) and *virtual private networks* (VPNs) that extend Ethernet's reach—add to the pressure to push the use of Ethernet transport.

In this section of the book we discuss metro players: who they are, what they do, and how they are positioned. As mentioned briefly earlier, the metro environment has three key segments: the component manufacturers that create semiconductor and opto-electronic devices; the systems manufacturers that buy components to assemble complex devices such as routers, switches, and multiplexers; and the service providers that buy systems to be used within their networks. We begin our discussion with the component manufacturers.

Figure 5-1
The metro marketplace "food chain"

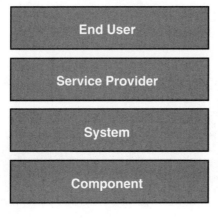

Component Manufacturers

Component manufacturers design, develop, and build both opto-electronic and semiconductor solutions for a wide array of applications. Opto-electronic components, for example, are used to transmit, process, change, amplify, and receive optical signals that carry voice and data traffic over in-place optical networks. Integrated circuits, on the other hand, also known as ICs or simply chips, perform a variety of functions, including processing electronic signals, controlling electronic system functions, and processing and storing data.

A growing trend exists among component manufacturers to increase the value of the devices they sell. One trend today is the so-called system on a chip, in which multiple are combined on a single silicon substrate to create a complex device that performs an array of functions. For example, multiple devices can be combined into single products, called modules, which deliver functionality to customers traditionally performed by a number of separate devices.

At the most fundamental level, systems comprise the four components shown in Figure 5-2:

- *Access interfaces*, which connect to access devices that generate and terminate traffic on the customer side of the box
- *Switch fabrics*, used to establish paths through the network
- *Processors and drivers*, used to control the switching process
- *Trunk interfaces*, which connect the box to the high-speed transport network

Component manufacturers design and develop all these devices and more, allowing system designers to create complex networking solutions from a collection of functional components.

Figure 5-2
The four functional components of a typical system-level device

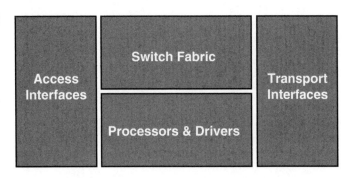

Consider the illustration shown in Figure 5-3. Here we see a *Dense Wavelength Division Multiplexing* (DWDM)-enhanced *Synchronous Optical Network / Synchronous Digital Hierarchy* (SONET/SDH) ring used to interconnect four *optical add-drop multiplexers* (OADMs). The OADMs, in turn, connect to SONET/SDH switches that in turn connect to multiservice switches or routers that provide multiprotocol connectivity and transport. The SONET/SDH switches connect to either traditional rings or to multichannel DWDM transport.

Let's now take a closer look at the devices shown in the preceding diagram (see Figure 5-4). Here we see a variety of components. On the left, we have DWDM transponders, which integrate multichannel optics and integrated circuits on a single functional module. On the right, we have a selection of components, including *Erbium-Doped Fiber Amplifier* (EDFA) modules, which help to extend the distance over which an optical signal can be transmitted; EDFA pump lasers, which create the signal that is to be transmitted; *dynamic gain equalization filters* (DGEFs), which help to improve the signal-to-noise ratio on transmitted signals and therefore the transmission distance; and a variety of passive devices, including thin-film filters, switches, attenuators, and isolators that increase the bandwidth of multichannel DWDM systems.

The SONET/SDH switch, shown in Figure 5-5, is potentially somewhat more complex in keeping with its high degree of functional responsibility

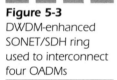

Figure 5-3
DWDM-enhanced
SONET/SDH ring
used to interconnect
four OADMs

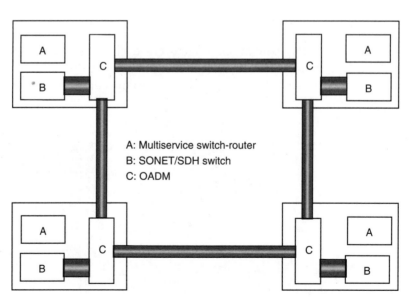

A: Multiservice switch-router
B: SONET/SDH switch
C: OADM

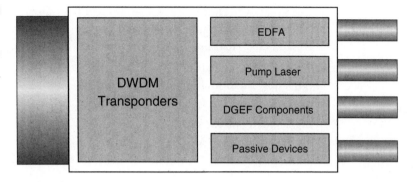

Figure 5-4
A more detailed look at the anatomy of the system

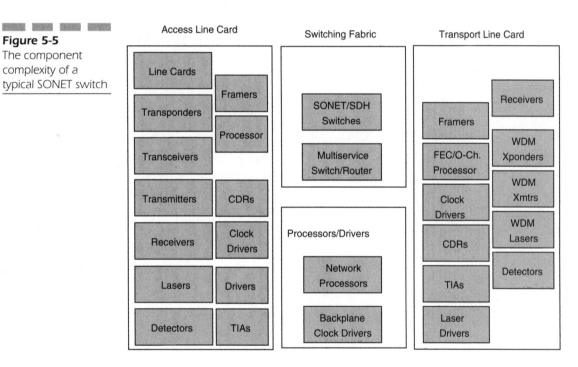

Figure 5-5
The component complexity of a typical SONET switch

within the network. This device has four functional areas: access functions, switching fabric functions, processor/driver functions, and trunk-side access functions.

The access side is shown here to connect to a SONET/SDH access ring. This functional region includes the following:

■ ADM line cards that integrate the full functionality of an ADM, including maintenance interfaces

- Transponders that integrate an optical transmitter and receiver into a single package that also includes clock signal synthesis and recovery as well as electrical multiplex-demultiplex functionality

- Transceivers that provide bidirectional transmission of Ethernet payloads, including Gigabit Ethernet

- Transmitters and receivers that respectively originate and terminate high-speed optical signals

- Lasers that create the high-quality optical signal that is to be transmitted

- Laser drivers that reduce signal jitter and skew, therefore producing a clean laser signal

- Clock drivers that ensure the accuracy of the clock signal

- *Clock data recovery* (CDR) used to ensure device synchronization

- Framers and processors that process frame structures and pointers in SONET/SDH environments

The switching fabric comprises SONET/SDH switches and multiservice switch-router assemblies, which include *Asynchronous Transfer Mode* (ATM), *Time Division Multiplexing* (TDM), and *Internet Protocol* (IP) fabrics.

The processors and drivers subassembly includes network processors and backplane clock drivers. Network processors offer a wide array of on-board functions, including layer two through seven protocol processing, buffer management, traffic shaping, data modification, and per-flow policing and statistics collection. These devices are often found in multiservice switch routers, *Digital Subscriber Line Access Multiplexers* (DSLAMs), metro access and core systems, and wireless base stations.

The trunk-side access functions are equally complex. Here we find framer processors, clock drivers, CDRs, lasers, laser drivers, transmitters, and receivers as before, but we also find a number of additional devices. *Forward Error Correction* (FEC) processors reduce the bit error rate on transmitted signals so that the overall bit rate (and therefore throughput) can be increased on the transmitted signal. WDM transponders, transmitters, and receivers are similar to those found on the access side of the device, but they are designed to handle the multichannel requirements of DWDM systems. Detectors do exactly what their name implies: They detect and receive an incoming optical signal prior to handing it off for processing. Finally, *transimpedance amplifiers* (TIAs) are used to perform signal conversion in environments where low noise and high signal gain are important.

The functions performed by the devices manufactured by component suppliers are marvelously complex. The devices described briefly here are but a few of the broad range of components offered by this sector of the marketplace. For a more complete listing of the devices that are available, visit the web site of any one of the companies shown in this chapter.

The Component Players

The major players in the component industry, arranged in order of revenue and marketshare, are the following:

- Agere
- Intel
- Texas Instruments
- Motorola
- STMicroelectronics
- Infineon
- JDS Uniphase
- Toshiba
- Fujitsu
- NEC
- QUALCOMM
- Phillips
- Mitsubishi
- Advanced Micro Devices
- Conexant
- Analog Devices
- Xilinx
- Hitachi
- Agilent
- Matsushita

Systems Manufacturers

It is, of course, close to impossible to create a complete list of all the companies that manufacture systems targeted at the metro marketplace. For purposes of representation, I have chosen a selection of them here that illustrates the broad range of the systems that they collectively offer.

Appian Communications

Appian started with the goal of creating a platform that could carry Ethernet over carrier SONET infrastructures. The company's primary *Optical Services Activation Platform* (OSAP) 4800 product enables a service provider's customers to connect to the existing metropolitan network via an Ethernet connection in the Appian box, a connection that would typically be located in a building or corporate campus. As demand increases, bandwidth can be added in increments as small as 64 Kbps to as much as Gigabit Ethernet. Users migrating from ATM or Frame Relay can expect the same *quality of service* (QoS) that those technologies provide.

OSAP enables a carrier to use Ethernet wherever it chooses while connecting to a variety of legacy technologies such as ATM and Frame Relay. OSAP has a hybrid packet and TDM platform that easily transports traditional voice, data, and packet traffic. Carriers can encapsulate Ethernet within SONET frames or convert them to Frame Relay frames or ATM cells.

Appian has recently announced the addition of multipoint services with its virtual Ethernet rings, logical rings that can be created to link disparate sites across a carrier's existing metro network. With a single OSAP node in each city, a carrier can extend this logical ring model across a national or international network (www.appiancom.com).

Caly Networks

CALY Networks, a startup based in Gerrards Cross, England, and Sunnyvale, California, offers a dynamic mesh wireless network architecture that supports low-cost broadband access. Their service model, called CALYnet, relies on wireless nodes that simultaneously serve as a customer access point and routing node. No base stations are used in the CALY model. Their "every-which-way" mesh overcomes line-of-sight issues and allows the network to be extended a node at a time.

Core elements of CALY's platform include *media access control* (MAC) layer software, *Space-Time Scheduled Multiple Access* (STS/MA) wireless routing algorithms, a high-speed modem, and intelligent beam-steering antenna technology. The system combines layers one through three routing and switching techniques (www.calynetworks.com).

Cisco

It should come as no surprise that Cisco has a major presence in the metro domain. The company recently announced a four-slot version of its 12000 series router designed to extend IP from core *Internet service provider* (ISP) *points of presence* (POPs) to the metro and edge, all at OC-192.

Cisco is aggressively marketing its own version of the emerging *Resilient Packet Ring* (RPR) standard. The 10720 Internet router, a low-cost access router, supports *dynamic packet transport* (DPT), Cisco's proprietary version of RPR. The new router fits into an existing line of 10 Gbps products that includes the Cisco 12416, a 16-slot chassis that fits into a 7-foot telco rack; the Cisco 12410, a 10-slot chassis that fits into half a rack; and the 12406, a six-slot chassis that takes up one-quarter of a standard telco rack.

Cisco also offers the 12404, which sports an 80 Gbps switch fabric and occupies 5 rack units. The 12404 supports all existing 12000 series line cards up and including OC-192 as well as the company's relatively new IP services engine line cards. These cards facilitate high-speed services, including QoS, multicast, and security.

The company also has a presence in the metro DWDM space. Their recently announced metro DWDM platform aggregates as many as 40 *Enterprise System Connection* (ESCON) data connections over a single 10 Gbps channel. The ONS 15530 carries up to 32 wavelengths at either 2.5 Gbps or 10 Gbps; supports SONET/SDH, ATM, and *Fibre Channel Connectivity* (FICON) protocols; and provides DWDM features such as a management- and performance-monitoring optical supervisory channel and a line of four-channel OADM modules. The platform can be configured in a variety of ways, including point-to-point, ring, or meshed topologies.

Extreme Networks

Extreme Networks (www.extremenetworks.com) targets the lucrative greenfield Ethernet metro service provider. These alternative service providers have the unique opportunity of implementing Ethernet without worrying about interworking requirements with existing networks and without redesigning existing business practices and operational processes to support a new Ethernet service business.

Existing service providers have invested heavily in ATM, Frame Relay, and other circuit-based technologies because they work well for both WANs and *metro area networks* (MANs). This means that incumbent providers must have a seamless migration strategy that takes advantage of both their investment in legacy equipment and newly installed leading-edge technology. But because their networks were originally designed to specifically support voice and private-line data service, they are not considered optimal for Internet traffic and high-speed data services.

Extreme Networks provides a migration strategy that bridges the gap between the existing infrastructure and new Ethernet-based architectures. This allows an incumbent provider to improve data transport efficiency without disrupting the revenue stream from a legacy network or compromising availability, scalability, and revenue generation.

Extreme Networks also offers highly flexible bandwidth-provisioning features. These include a comprehensive set of QoS features for Ethernet switches and enable wire-speed traffic classification, standards-based marking, and expedited forwarding over eight queues per port.

The Extreme *Infrastructure and Services Management* (ISM) comprehensive suite of provisioning, monitoring, and service-level enforcement tools supports the ability of a service provider to design and deliver network services by automating service creation, activation, delivery, and billing.

Foundry Networks

Foundry's solutions are targeted to take advantage of the convergence of IP, optical, and Ethernet technologies to bring the best of each to the MAN. The company's products combine Ethernet, SONET, and DWDM to offer a solution for broadband IP infrastructure, turning bandwidth into profitable revenue streams (www.foundrynetworks.com).

Lucent Technologies

Lucent's metro product line is built around the following customer networking issues:

- Skyrocketing customer demand for capacity and new services
- Changing traffic patterns

- Multiprotocol environments
- The need for greater revenue and profits

The company's Metropolis® products offer reliable, cost-effective, scalable transport for high-traffic networks. In keeping with the demands to ensure interoperability between legacy and new products, they work well with both new multiservice data switches and existing TDM and packet-based equipment to support next-generation SONET/SDH and the ongoing transition from TDM to data services. These include Gigabit Ethernet, IP, and Metro DWDM solutions for both access and core networking environments.

Products supported include

- Dedicated private-line services
- Internet access
- Ethernet access
- *Storage Area Networking* (SAN)
- Dedicated transparent wavelength services

The Metropolis DMX access multiplexers and Metropolis *Enhanced Optical Networking* (EON) DWDM system increase performance and reduce operating expenses. Lucent's innovative Navis™ Optical Management Solution software works with these optical products to automatically discover available network elements, assess network capacity, and perform automatic configuration for best routes. Similarly, the LambdaUnite™ *MultiService Switch* (MSS) facilitates a smooth migration from ring to mesh topologies. It can serve as a hub and cross-connect to bridge the metro and core with a single product to reduce operating and capital expenses. Readers will recall that earlier in the book we made the point that optical products have the capability to reduce capital expenditures by reducing the total number of network elements required to build and operate a network.

LuxN

LuxN has historically made optical transport products for the access region of the network. With the growth in metro interest, they now offer products designed to interconnect regional fiber rings.

LuxN's WS 6400, which improves upon the job done by its WS 3200 predecessor, combines optical transport and wavelength switching in a single device. It offers 64 wavelengths over a single-fiber pair in a single shelf, as opposed to the 32 wavelengths provided by the WS 3200. It also improves upon the WS 3200 in another way. Unlike the WS 3200, which is a point-to-point DWDM transport solution, the WS 6400 combines DWDM with lambda switching, which means that service providers can connect multiple fiber rings within a region to create a meshed network.

Riverstone Networks

Riverstone Networks, Inc. (www.riverstonenetworks.com) builds routers that, in the words of the company, "convert raw bandwidth into profitable services for Metropolitan Area Networks." Riverstone's products support the delivery of metro services over a range of access technologies, including SONET/SDH, Gigabit Ethernet, T1/E1/T3/E3, and DWDM. Riverstone also offers granular, connection-oriented usage accounting, an important component of any well-managed network.

Riverstone has customers in more than 50 countries and in some of the world's largest networks, including Cox Communication, British Telecom, Qwest, and Telefónica. The company's product line includes the RS router family designed for aggregation, access, and data hosting.

Sycamore Networks

Sycamore (www.sycamorenetworks.com) offers a wide range of Ethernet service solutions, including Ethernet and optical switching technologies that take advantage of the existing infrastructure of incumbent carriers but that want to offer a new class of service. By offering intelligent optical switching, Sycamore offers carriers the ability to deliver enhanced scalability, capacity utilization, and flexible protection for carrier-class performance while still offering a variety of Ethernet services.

Sycamore Networks' Ethernet Service Solutions deliver Ethernet and software-rich optical switching while supporting legacy SONET/SDH network standards and protocols. Their four-port GbE Service Module for the SN 3000 Intelligent Optical Switch enables carriers to use their existing infrastructure to deliver highly reliable Gigabit Ethernet services, including transparent *local area networks* (LANs), Ethernet private lines, and data aggregation-to-optical access for backbone routers/switches. Sycamore solutions rely on standards-based virtual concatenation for efficient bandwidth utilization and advanced protection schemes for flexible service offerings.

Service provider benefits from Sycamore include sub- to full-rate Gigabit Ethernet service delivery capabilities, rapid Ethernet service provisioning, reduced network complexity, efficient capacity utilization, simplified network-wide capacity planning and network flexibility, carrier-class reliability and manageability, high density, small footprint Gigabit Ethernet (up to 120 ports per bay), and support for a broad range of protection schemes.

End users, on the other hand, benefit from flexible capacity deployment, highly trained technical support personnel, rapid and efficient provisioning, low-cost data interfaces (compared to SONET/SDH), and the ability to enjoy carrier-class Gigabit Ethernet services.

Tellabs

Tellabs (www.tellabs.com) prides itself on building highly scalable solutions for the metro network that can be adapted easily to changing traffic patterns. The Tellabs® 6410 transport edge node is an optical access unit that transports both SONET and Ethernet and is ideal for low-cost access. As a combined solution, the Tellabs 6400 transport switch and the Tellabs 7100 optical transport system enable a service provider to provision Gigabit Ethernet, IP, and Frame Relay while also supporting SAN solutions such as ESCON, FICON, and Fibre Channel.

Clearly, metro systems have many more vendors, but this will provide a sense of the variety of the players that are in the market today. We now turn our attention to the service providers.

Service Provider Issues

As service providers deploy MANs today, they face not only a great deal of market uncertainty, but also unprecedented opportunities to enter new markets with a suite of new products. LANs are faltering under the inexorable demand for bandwidth, and corporate *information technology* (IT) personnel are scrambling to keep up with it. The situation is not helped by the fact that applications are emerging that demand the bandwidth, and the costs of both bandwidth and networking equipment are plummeting. In parallel to this onslaught, DWDM technology has come booming into the metro space, further drawing down the expected cost of bandwidth.

The good news is that companies positioned to do so (metro service providers) can now bridge the historical functionality gap that has always existed between LANs and the backbone networks that interconnect them. The difference is that today this interconnection can be down with high-bandwidth, low-cost, and highly flexible solutions such as Ethernet instead of legacy TDM solutions such as T1 and SONET.

At the same time, optical fiber has been installed in unprecedented quantities in metro environments, as service providers have anticipated the expected high demand for high-speed transport. As customers have become more and more aware of the capabilities of the new metro network and of the applications that will most likely reside there, they have pressured service providers to ensure that new metro designs include a variety of functional characteristics. These include

- *Scalability*, which boils down to the capability to scale from low-speed transport to high-speed Gigabit Ethernet and beyond

- *Low cost per transported bit,* in keeping with the plummeting cost of delivered bandwidth

- *Support for a range of so-called next-generation services,* including variable QoS, multiple physical interfaces, network management, effective billing, and IP support

- *Support for legacy voice transport* because voice may only represent a fraction of the total transported traffic volume, but it still accounts for well over 85 percent of total network revenues

- *Very high survivability* that mimics carrier-class operations with up-times in excess of 99.999 percent, including redundant hardware components and survivable optical ring architectures

As service providers build their networks, they rely on three types of systems to do so. These include SONET/SDH *Multiservice Provisioning Platforms* (MSPPs), layer 2/3 packet switches, and metro DWDM systems. Because the legacy network was primarily designed around the transport requirements of voice, it has a hard time adapting to the more flexible and unpredictably bursty performance characteristics of IP traffic, which is rapidly becoming the preferred transport protocol in today's networks. As the network is called upon to transport more and more multiprotocol traffic, a new type of device is called for. These devices, known as MSPPs, map a wide variety of traffic types into SONET payloads for transport across the legacy network infrastructure. Some of these MSPPs support layer two and layer three switching functions, but they implement them as functional layers above the physical layer. Functionally, they are nothing more than switches and routers working closely with SONET (or SDH) ADMs. The "Component Manufacturers" section previously discussed the integration of components to create a system on a chip; this is a macro example of that evolution.

A second type of MAN product relies on native-mode, packet-switched solutions, based on switches or routers, connected by point-to-point Ethernet links, packet-over-SONET facilities, or some other high-speed protocol. Some of these solutions (particularly those that rely on high-speed versions of Ethernet) offer efficient bandwidth utilization and sophisticated handling of multiple data services.

The downside of this solution is that they do not do a particularly good job of providing the QoS and survivability typically required for critical payloads because they do not guarantee end-to-end packet delivery. Pure Gigabit Ethernet options do not offer the option to offer both data and high-quality circuit services from a single converged platform, which forces the service provider to deploy multiple boxes in the network to accommodate customer requirements, a step that runs contrary to our argument about reducing the complexity of the network. Because these solutions come from the data side of the house, they lack the redundancy and robustness required to guarantee the carrier-class service levels that users of the public network have come to expect. Furthermore, Gigabit Ethernet does not yet adapt well to the typical metro ring topologies and does not typically offer 50-millisecond restoral. Newly released protocols such as RPR will overcome this problem.

The third system group adapts backbone optical transport devices for use in the metro. Metro DWDM transport systems, for example, offer a 32:1 transport volume improvement over a single fiber by mapping multiple signals into different wavelengths. Most of them rely on ring protection switching and can transport a variety of different traffic types (DWDM is protocol independent).

The downside to this family of devices is that metro DWDM systems typically map an input stream into its own wavelength, but they do not offer an efficient or flexible technique for aggregating lower bandwidth flows into DWDM. Furthermore, wavelength-level connectivity is difficult to provision because it is dependent upon fixed-wavelength optical components. As the components evolve, however, this limitation is being eliminated; multiple wavelength devices have now entered the market and will become more common over time.

So what should a modern metro network look like in terms of characteristics and functions? First, they should be designed around the idea that IP will be the protocol of choice in the very near future for the transport of most data types. This is a good idea, because it allows for a flexible service creation environment and therefore allows the service provider to control their deployment costs.

Second, the network should be packet-based, in keeping with its IP underpinnings. Packet transport is the most flexible and cost-effective mechanism for the transport of multiple data types and, with the onslaught of demand for new services, is the best choice.

Third, although the network should be built around an IP construct, it must also support legacy TDM-based services because of the enormous investment in these technologies that service providers have already made.

Fourth, the network must be overlaid with effective *Operations, Administration, Maintenance, and Provisioning* (OAM&P) and a network management/*operation support systems* (OSS) system to ensure that it can be properly billed, provisioned, monitored, maintained, and repaired. On to the service providers.

BellSouth

Bellsouth (www.bellsouth.com) recently announced that it will offer Gigabit Ethernet, storage services, and voice-enabled services. Like all the *incumbent local exchange carriers* (ILECs), BellSouth recognizes the importance of having a metro play and has made it clear that it intends to enter the game.

Broadwing

Broadwing (www.broadwing.com) was created from the merger of IXC Communications and Cincinnati Bell, an interesting combination of old and new, to be sure. Broadwing was the first service provider to offer a fully deployed, all-optical network. The company's expertise lies in broadband, Internet, back office, network management, and IT consulting arenas; they also own and operate the first national, next-generation IP network.

As far as network quality is concerned, Broadwing does well. Their network offers 100 percent IP availability, 99 percent or better packet delivery, and 50-millisecond roundtrip latency in the United States. Using next-generation fiber, DWDM, and Corvis™ optical switching technology, the company offers expandable capacity for future customer demands. They also have a presence in the metro, with service in well over 90 percent of the major cities. Broadwing offers ATM, Frame Relay, Dial IP, dedicated Internet, private line, and *symmetric digital subscriber line* (SDSL) protocol support, and its network is enhanced with dual media diverse routing, hardware redundancy, a fully meshed, intelligently connected design, and duplicate peering connections.

Broadwing and Telseon recently formed a strategic alliance to offer end-to-end Gigabit Ethernet services across their networks. The arrangement provides Broadwing with additional metro connectivity, while Telseon will be able to offer its customers metro-to-metro connectivity. The services are currently in trial in a number of cities, including New York, San Francisco,

Los Angeles, Chicago, Dallas, Washington, D.C., Atlanta, Seattle, Miami, and Denver.

CityNet Communications

In the list of innovative services, CityNet (www.citynettelecom.com) has to be close to the top if not at the top. Instead of digging trenches to lay fiber optic cable, CityNet runs its fiber through city sewers. The company relies on a fleet of cylinder-shaped robots that string fiber inside of sewer mains. From a survivability point of view, this approach is valid, because sewer pipes are below other utility facilities and are rarely disrupted by backhoes.

IntelliSpace

IntelliSpace (www.intellispace.com) is an Ethernet carrier that provides in-building Ethernet services. The company recently secured $70 million in new funding for the continuing rollout of its network within its existing markets. So far, the company has provided services in 1,000 commercial buildings, including Rockefeller Center, Lincoln Center, and the Chicago Board of Options Exchange at One Financial Plaza. The carrier also has a significant contract in the United Kingdom through Land Securities, providing it with access to 350 commercial properties.

OnFiber Communications

OnFiber Communications Inc. (www.onfiber.com) is constructing fiber optic local access networks in promising business corridors within major U.S. metro areas. The company's goal is to resolve the famous (perhaps infamous?) last-mile bottleneck. By providing end-to-end optical connectivity between service providers and their customers, OnFiber's network will enable other service providers to take advantage of a wider variety of high-demand broadband applications. The company is a startup and is currently financed by Kleiner Perkins Caufield & Byers, Incepta (an affiliate of Bechtel Enterprises), Bear Stearns Merchant Banking, Amerindo, Telesoft Partners, General Electric Capital Corporation, and numerous other leading investment firms.

OnFiber's network is protocol and bit-rate independent and supports dedicated SONET, switched SONET, optical Ethernet, and optical wavelength services over a single network. OnFiber also offers a customer-centric OSS that enables the company to provision circuits in a matter of weeks instead of months. Its pricing is also attractive; OnFiber offers services at prices that are significantly lower than those of incumbent providers. Furthermore, the company's network management system enables customers to buy incremental bandwidth, thus controlling their costs with significant granularity.

SBC

SBC Communications (www.sbc.com) offers the broadest metro service package of any ILEC. The company recently its *Multiservice Optical Networking* (MON) product, which provides dedicated point-to-point service at 2.5 Gbps per wavelength between data centers or business sites in a metro area. MON carries as many as 160 Gbps of unprotected traffic or 80 Gbps of protected traffic on a single fiber. It supports numerous transport protocols, including ESCON, FICON, *Early Token Release* (ETR), *International Standards Commission* (ISC), Fibre Channel, Fast Ethernet, Gigabit Ethernet, SONET, and D1 video in their native formats, keeping with the companies strong focus on customer applications and the need to match technologies to customer requirements.

Storm Telecommunications

Storm Telecommunications (www.stormtel.com) offers metro Gigabit Ethernet services in London, Paris, Amsterdam, Hamburg, Stuttgart, and Frankfurt. The company is strategically positioning its service as an alternative to SDH.

Telseon

Telseon (www.telseon.com) offers both metro Ethernet services, scalable from 1 Mbps to 1 Gbps, as well as metro wavelength services up to 10 Gbps. The company has networks in 20 major markets and boasts more than 120 points of presence.

Telseon recently endorsed free-space optics, the laser-based wireless technology that links buildings in a campus environment. The company also recently added OC-12 services to its OC-48 and OC-192 flexible wavelength services. The OC-12 service is targeted at companies looking for smaller bandwidth increments.

Williams Communications

Williams Communications (www.williamscommunications.com) offers its metro access services via 80 metro access points in 17 U.S. cities. The company has provisioned fiber from POPs along its nationwide network to a variety of strategic aggregation facilities, including carrier hotels, collocation providers, central offices, network access points, and Internet data centers. The service includes a DS-3 through an OC-48 local private line and an OC-48 lambda service.

XO Communications

XO Communications (www.xo.com) offers OC-12, OC-48, and OC-192 metro wavelength services in 62 U.S. markets. The company's XO *Metro Wavelength Service* (MWS) is a fully protected protocol-transparent service targeted at both carrier and enterprise customers and is designed to carry traffic between multiple sites in a metro region at a fixed monthly cost.

Yipes

Yipes (www.yipes.com) provides scalable business-to-business and business-to-Internet IP networking services. The company's regional IP network, Yipes MAN, seamlessly extends a customer's LAN to other business locations.

Yipes offers an impressive array of services. Yipes NET Internet service provides LAN connectivity to the ISP, while Yipes WALL provides a managed firewall service. Yipes *National Area Network* (NAN) replaces private-line and Frame Relay services with Ethernet. Finally, Yipes TOUGH enables customers to augment network redundancy between their premises and the Yipes network.

Yipes currently has 300 buildings in service and has more than 32,000 fiber miles in 128 rings. Fiber route miles currently stand at 3,600. The company offers services in 21 markets across the country.

Conclusion

The arrival of optical networking, particularly in the metro network, promises improved network efficiencies that permit service providers to deliver a wide array of services to their customers with significant efficiency. They will be able to offer *quality of service* (QoS) guarantees for their intelligent optical networks and *service level agreements* (SLAs) that support those services and enable them to charge variable fees for various service types. For service providers to deploy end-to-end, all-optical networks, however, they *must* deploy optical technology in the metro domain. By implementing *metropolitan optical networks* (MONs), service providers will be able to offer the benefits of optical networking not only to their core customers, but also to their end customers. These benefits offer dynamic, standards-based multiprotocol support as well as cost-effective provisioning, survivability, and integrated network management. By adopting the MON concept, service providers will be able to build a truly future-proof network that will guarantee sustainable long-term competitive advantages.

The benefits of a metro network are legion and accrue not only to enterprise customers, but to component manufacturers, system builders, and service providers as well. By delivering intelligent optical bandwidth beyond the edge, service providers can deliver a wide array of enhanced capabilities that will help them climb the food chain and become much more than simply bandwidth providers. As they do so, their networks will evolve in capability, and as they succeed, so too will all levels of the networking hierarchy.

So should it be.

Common Industry Acronyms

AAL	ATM Adaptation Layer
AARP	AppleTalk Address Resolution Protocol
ABM	Asynchronous Balanced Mode
ABR	Available bit rate
AC	Alternating current
ACD	Automatic call distribution
ACELP	Algebraic Code-Excited Linear Prediction
ACF	Advanced communication function
ACK	Acknowledgment
ACM	Address complete message
ACSE	Association Control Service Element
ACTLU	Activate logical unit
ACTPU	Activate physical unit
ADCCP	Advanced Data Communications Control Procedures
ADM	Add-drop multiplexer
ADPCM	Adaptive Differential Pulse Code Modulation
ADSL	Asymmetric Digital Subscriber Line
AFI	Authority and Format Identifier
AIN	Advanced Intelligent Network
AIS	Alarm Indication Signal
ALU	Arithmetic logic unit
AM	Administrative Module (Lucent 5ESS); Amplitude Modulation
AMI	Alternate Mark Inversion
AMP	Administrative Module Processor
AMPS	Advanced Mobile Phone System
ANI	Automatic Number Identification (SS7)
ANSI	American National Standards Institute
APD	Avalanche Photodiode
API	Application programming interface
APPC	Advanced Program-to-Program Communication
APPN	Advanced Peer-to-Peer Networking
APS	Automatic Protection Switching
ARE	All Routes Explorer (Source Route Bridging)

ARM	Asynchronous Response Mode
ARP	Address Resolution Protocol (IETF)
ARPA	Advanced Research Projects Agency
ARPANET	Advanced Research Projects Agency Network
ARQ	Automatic Repeat Request
ASCII	American Standard Code for Information Interchange
ASI	Alternate Space Inversion
ASIC	Application-Specific Integrated Circuit
ASK	Amplitude Shift Keying
ASN	Abstract Syntax Notation
ASP	Application service provider
AT&T	American Telephone and Telegraph
ATDM	Asynchronous Time Division Multiplexing
ATM	Asynchronous Transfer Mode; Automatic Teller Machine
ATMF	ATM Forum
AU	Administrative Unit (SDH)
AUG	Administrative Unit Group (SDH)
AWG	American Wire Gauge
B8ZS	Binary 8 Zero Substitution
BANCS	Bell Administrative Network Communications System
BBN	Bolt, Beranak, and Newman
BBS	Bulletin Board Service
Bc	Committed Burst Size
BCC	Blocked Calls Cleared; Block Check Character
BCD	Blocked Calls Delayed
BCDIC	Binary Coded Decimal Interchange Code
Be	Excess Burst Size
BECN	Backward Explicit Congestion Notification
BER	Bit Error Rate
BERT	Bit Error Rate Test
BGP	Border Gateway Protocol (IETF)
BIB	Backward Indicator Bit (SS7)
B-ICI	Broadband Intercarrier Interface

BIOS	Basic Input/Output System
BIP	Bit interleaved parity
B-ISDN	Broadband Integrated Services Digital Network
BISYNC	Binary Synchronous Communications Protocol
BITNET	Because It's Time Network
BITS	Building Integrated Timing Supply
BLSR	Bidirectional line switched ring
BOC	Bell Operating Company
BPRZ	Bipolar Return to Zero
Bps	Bits per second
BRI	Basic rate interface
BRITE	Basic rate interface transmission equipment
BSC	Binary synchronous communications
BSN	Backward sequence number (SS7)
BSRF	Bell System Reference Frequency
BTAM	Basic Telecommunications Access Method
BUS	Broadcast unknown server
CAD	Computer-Aided Design
CAE	Computer-Aided Engineering
CAM	Computer-Aided Manufacturing
CAP	Carrierless amplitude/phase modulation; Competitive Access Provider
CARICOM	Caribbean Community and Common Market
CASE	Common Application Service Element; Computer-Aided Software Engineering
CAT	Computer-Aided Tomography
CATIA	Computer-Assisted Three-dimensional Interactive Application
CATV	Community antenna television
CBEMA	Computer and Business Equipment Manufacturers Association
CBR	Constant bit rate
CBT	Computer-based training
CC	Cluster controller

CCIR	International Radio Consultative Committee
CCIS	Common Channel Interoffice Signaling
CCITT	International Telegraph and Telephone Consultative Committee
CCS	Common Channel Signaling; hundred call seconds per hour
CD	Collision detection; compact disc
CDC	Control Data Corporation
CDMA	Code Division Multiple Access
CDPD	Cellular Digital Packet Data
CD-ROM	Compact Disc-Read Only Memory
CDVT	Cell Delay Variation Tolerance
CEI	Comparably Efficient Interconnection
CEPT	Conference of European Postal and Telecommunications Administrations
CERN	European Council for Nuclear Research
CERT	Computer Emergency Response Team
CES	Circuit Emulation Service
CEV	Controlled Environmental Vault
CGI	Common gateway interface (Internet)
CHAP	Challenge Handshake Authentication Protocol
CICS	Customer Information Control System
CICS/VS	Customer Information Control System/Virtual Storage
CIDR	Classless Interdomain Routing (IETF)
CIF	Cells in frames
CIR	Committed information rate
CISC	Complex Instruction Set Computer
CIX	Commercial Internet Exchange
CLASS	Custom Local Area Signaling Services (Bellcore)
CLEC	Competitive Local Exchange Carrier
CLLM	Consolidated Link Layer Management
CLNP	Connectionless Network Protocol
CLNS	Connectionless Network Service
CLP	Cell Loss Priority

CM	Communications Module (Lucent 5ESS)
CMIP	Common Management Information Protocol
CMISE	Common Management Information Service Element
CMOL	CMIP Over LLC
CMOS	Complementary Metal Oxide Semiconductor
CMOT	CMIP Over TCP/IP
CMP	Communications Module Processor
CNE	Certified NetWare engineer
CNM	Customer Network Management
CNR	Carrier-to-noise ratio
CO	Central office
CoCOM	Coordinating Committee on Export Controls
CODEC	Coder-decoder
COMC	Communications controller
CONS	Connection-Oriented Network Service
CORBA	Common Object Request Brokered Architecture
COS	Class of service (APPN); Corporation for Open Systems
CPE	Customer premises equipment
CPU	Central processing unit
C/R	Command/response
CRC	Cyclic redundancy check
CRM	Customer Relationship Management
CRT	Cathode ray tube
CRV	Call reference value
CS	Convergence sublayer
CSA	Carrier serving area
CSMA	Carrier Sense Multiple Access
CSMA/CA	Carrier Sense Multiple Access with Collision Avoidance
CSMA/CD	Carrier Sense Multiple Access with Collision Detection
CSU	Channel service unit
CTI	Computer Telephony Integration
CTIA	Cellular Telecommunications Industry Association
CTS	Clear to send

CU	Control unit
CVSD	Continuously Variable Slope Delta Modulation
CWDM	Coarse Wavelength Division Multiplexing
D/A	Digital-to-analog
DA	Destination address
DAC	Dual attachment concentrator (FDDI)
DACS	Digital Access and Cross-connect System
DARPA	Defense Advanced Research Projects Agency
DAS	Dual-attachment station (FDDI); direct attached storage
DASD	Direct access storage device
DB	Decibel
DBS	Direct broadcast satellite
DC	Direct current
DCC	Data Communications Channel (SONET)
DCE	Data circuit-terminating equipment
DCN	Data Communications Network
DCS	Digital Cross-connect System
DCT	Discrete Cosine Transform
DDCMP	Digital Data Communications Management Protocol (DNA)
DDD	Direct distance dialing
DDP	Datagram Delivery Protocol
DDS	DATAPHONE Digital Service or Digital Data Service
DE	Discard eligibility (LAPF)
DECT	Digital European Cordless Telephone
DES	Data Encryption Standard (NIST)
DID	Direct inward dialing
DIP	Dual inline package
DLC	Digital loop carrier
DLCI	Data link connection identifier
DLE	Data Link Escape
DLSw	Data Link Switching
DM	Delta Modulation; Disconnected Mode; Data mining
DMA	Direct memory access (computers)

DMAC	Direct Memory Access Control
DME	Distributed Management Environment
DMS	Digital multiplex switch
DMT	Discrete multitone
DNA	Digital Network Architecture
DNIC	Data Network Identification Code (X.121)
DNIS	Dialed Number Identification Service
DNS	Domain Name System (IETF)
DOD	Direct outward dialing; Department of Defense
DOJ	Department of Justice
DOV	Data Over Voice
DPSK	Differential Phase Shift Keying
DQDB	Distributed Queue Dual Bus
DRAM	Dynamic Random Access Memory
DSAP	Destination Service Access Point
DSF	Dispersion-shifted fiber
DSI	Digital Speech Interpolation
DSL	Digital Subscriber Line
DSLAM	Digital Subscriber Line Access Multiplexer
DSP	Digital Signal Processing
DSR	Data Set Ready
DSS	Digital Satellite System; Digital Subscriber Signaling System
DSSS	Direct Sequence Spread Spectrum
DSU	Data service unit
DTE	Data terminal equipment
DTMF	Dual tone multifrequency
DTR	Data Terminal Ready
DVRN	Dense Virtual Routed Networking (Crescent)
DWDM	Dense Wavelength Division Multiplexing
DXI	Data exchange interface
EBCDIC	Extended Binary Coded Decimal Interchange Code
ECMA	European Computer Manufacturer Association
ECN	Explicit Congestion Notification

ECSA	Exchange Carriers Standards Association
EDFA	Erbium-Doped Fiber Amplifier
EDI	Electronic Data Interchange
EDIBANX	EDI Bank Alliance Network Exchange
EDIFACT	Electronic Data Interchange For Administration, Commerce, and Trade (ANSI)
EFCI	Explicit Forward Congestion Indicator
EFTA	European Free Trade Association
EGP	Exterior Gateway Protocol (IETF)
EIA	Electronics Industry Association
EIGRP	Enhanced Interior Gateway Routing Protocol
EIR	Excess information rate
EMBARC	Electronic Mail Broadcast to a Roaming Computer
EMI	Electromagnetic interference
EMS	Element Management System
EN	End node
ENIAC	Electronic Numerical Integrator and Computer
E/O	Electrical-to-optical
EO	End office
EOC	Embedded Operations Channel (SONET)
EOT	End of transmission (BISYNC)
EPROM	Erasable Programmable Read-Only Memory
ERP	Enterprise resource planning
ESCON	Enterprise System Connection (IBM)
ESF	Extended Superframe Format
ESP	Enhanced service provider
ESS	Electronic Switching System
ETSI	European Telecommunications Standards Institute
ETX	End of text (BISYNC)
EWOS	European Workshop for Open Systems
FACTR	Fujitsu Access and Transport System
FAQ	Frequently Asked Questions
FAT	File Allocation Table

FCS	Frame Check Sequence
FDD	Frequency Division Duplex
FDDI	Fiber Distributed Data Interface
FDM	Frequency Division Multiplexing
FDMA	Frequency Division Multiple Access
FDX	Full-duplex
FEBE	Far End Block Error (SONET)
FEC	Forward error correction; Forward Equivalence Class
FECN	Forward Explicit Congestion Notification
FEP	Front-end processor
FERF	Far End Receive Failure (SONET)
FET	Field Effect Transistor
FHSS	Frequency Hopping Spread Spectrum
FIB	Forward Indicator Bit (SS7)
FIFO	First In, First Out
FITL	Fiber in the Loop
FLAG	Fiber Ling Across the Globe
FM	Frequency Modulation
FOIRL	Fiber Optic Interrepeater Link
FPGA	Field Programmable Gate Array
FR	Frame Relay
FRAD	Frame Relay Access Device
FRBS	Frame Relay Bearer Service
FSK	Frequency Shift Keying
FSN	Forward Sequence Number (SS7)
FTAM	File Transfer, Access, and Management
FTP	File Transfer Protocol (IETF)
FTTC	Fiber to the Curb
FTTH	Fiber to the Home
FUNI	Frame User-to-Network Interface
FWM	Four Wave Mixing
GATT	General Agreement on Tariffs and Trade
GbE	Gigabit Ethernet

Gbps	Gigabits per second (billion bits per second)
GDMO	Guidelines for the Development of Managed Objects
GEOS	Geosynchronous earth orbit satellites
GFC	Generic Flow Control (ATM)
GFI	General Format Identifier (X.25)
GFP	Generic Framing Procedure
GFP-F	Generic Framing Procedure—Frame-Based
GFP-X	Generic Framing Procedure—Transparent
GMPLS	Generalized MPLS
GOSIP	Government Open Systems Interconnection Profile
GPS	Global Positioning System
GRIN	Graded index (fiber)
GSM	Global System for Mobile Communications
GUI	Graphical User Interface
HDB3	High Density, Bipolar 3 (E-Carrier)
HDLC	High-level Data Link Control
HDSL	High-bit-rate Digital Subscriber Line
HDTV	High-definition television
HDX	Half-duplex
HEC	Header Error Control (ATM)
HFC	Hybrid fiber/coax
HFS	Hierarchical File Storage
HLR	Home Location Register
HPPI	High Performance Parallel Interface
HSSI	High-Speed Serial Interface (ANSI)
HTML	Hypertext Markup Language
HTTP	Hypertext Transfer Protocol (IETF)
HTU	HDSL transmission unit
I	Intrapictures
IAB	Internet Architecture Board (formerly Internet Activities Board)
IACS	Integrated Access and Cross-connect System
IAD	Integrated Access Device

IAM	Initial Address Message (SS7)
IANA	Internet Address Naming Authority
ICMP	Internet Control Message Protocol (IETF)
IDP	Internet Datagram Protocol
IEC	Interexchange carrier (also IXC); International Electrotechnical Commission
IEEE	Institute of Electrical and Electronics Engineers
IETF	Internet Engineering Task Force
IFRB	International Frequency Registration Board
IGP	Interior Gateway Protocol (IETF)
IGRP	Interior Gateway Routing Protocol
ILEC	Incumbent local exchange carrier
IML	Initial microcode load
IMP	Interface Message Processor (ARPANET)
IMS	Information Management System
InARP	Inverse Address Resolution Protocol (IETF)
InATMARP	Inverse ATMARP
INMARSAT	International Maritime Satellite Organization
INP	Internet nodal processor
InterNIC	Internet Network Information Center
IP	Internet Protocol (IETF)
IPX	Internetwork Packet Exchange (NetWare)
IS	Information Systems
ISDN	Integrated Services Digital Network
ISO	International Organization for Standardization; Information Systems Organization
ISOC	Internet Society
ISP	Internet service provider
ISUP	ISDN user part (SS7)
IT	Information technology
ITU	International Telecommunication Union
ITU-R	International Telecommunication Union-Radio Communication Sector

IVD	Inside vapor deposition
IVR	Interactive Voice Response
IXC	Interexchange carrier
JEPI	Joint Electronic Paynets Initiative
JES	Job Entry System
JIT	Just in Time
JPEG	Joint Photographic Experts Group
KB	Kilobytes
Kbps	Kilobits per second (thousand bits per second)
KLTN	Potassium Lithium Tantalate Niobate
KM	Knowledge management
LAN	Local area network
LANE	LAN emulation
LAP	Link Access Procedure (X.25)
LAPB	Link Access Procedure Balanced (X.25)
LAPD	Link Access Procedure for the D-Channel
LAPF	Link Access Procedure to Frame Mode Bearer Services
LAPF-Core	Core Aspects of the Link Access Procedure to Frame Mode Bearer Services
LAPM	Link Access Procedure for Modems
LAPX	Link Access Procedure Half-duplex
LASER	Light Amplification by the Stimulated Emission of Radiation
LATA	Local Access and Transport Area
LCD	Liquid crystal display
LCGN	Logical channel group number
LCM	Line Concentrator Module
LCN	Local communications network
LD	Laser diode
LDAP	Lightweight Directory Access Protocol (X.500)
LEAF®	Large Effective Area Fiber® (Corning product)
LEC	Local exchange carrier
LED	Light-emitting diode

LENS	Lightwave Efficient Network Solution (Centerpoint)
LEOS	Low earth orbit satellites
LER	Label edge router
LI	Length indicator
LIDB	Line Information Database
LIFO	Last In, First Out
LIS	Logical IP subnet
LLC	Logical Link Control
LMDS	Local Multipoint Distribution System
LMI	Local Management Interface
LMOS	Loop Maintenance Operations System
LORAN	Long-range Radio Navigation
LPC	Linear Predictive Coding
LPP	Lightweight Presentation Protocol
LRC	Longitudinal Redundancy Check (BISYNC)
LS	Link state
LSI	Large Scale Integration
LSP	Label switched path
LSR	Label switched router
LU	Line unit; logical unit (SNA)
MAC	Media access control
MAN	Metropolitan area network
MAP	Manufacturing Automation Protocol
MAU	Medium attachment unit (Ethernet); multistation access unit (Token Ring)
MB	Megabytes
MBA™	Metro Business Access™ (Ocular)
Mbps	Megabits per second (million bits per second)
MD	Message Digest (MD2, MD4, MD5) (IETF)
MDF	Main Distribution Frame
MDU	Multidwelling unit
MEMS	Micro-electrical Mechanical System
MF	Multifrequency

MFJ	Modified Final Judgment
MHS	Message Handling System (X.400)
MIB	Management Information Base
MIC	Medium Interface Connector (FDDI)
MIME	Multipurpose Internet Mail Extensions (IETF)
MIPS	Millions of instructions per second
MIS	Management information systems
MITI	Ministry of International Trade and Industry (Japan)
ML-PPP	Multilink Point-to-Point Protocol
MMDS	Multichannel, Multipoint Distribution System
MMF	Multimode fiber
MNP	Microcom Networking Protocol
MON	Metropolitan optical network
MP	Multilink PPP
MPEG	Motion Picture Experts Group
MPLS	Multiprotocol Label Switching
MPλS	Multiprotocol Lambda Switching
MPOA	Multiprotocol over ATM
MRI	Magnetic Resonance Imaging
MSB	Most significant bit
MSC	Mobile Switching Center
MSO	Mobile Switching Office
MSPP	Multiservice provisioning platform
MSVC	Meta-signaling virtual channel
MTA	Major Trading Area
MTBF	Mean Time Between Failure
MTP	Message Transfer Part (SS7)
MTSO	Mobile Telephone Switching Office
MTTR	Mean Time to Repair
MTU	Maximum transmission unit; multitenant unit
MVS	Multiple virtual storage
NAFTA	North American Free Trade Agreement
NAK	Negative Acknowledgment (BISYNC, DDCMP)

NAP	Network Access Point (Internet)
NARUC	National Association of Regulatory Utility Commissioners
NAS	Network-attached storage
NASA	National Aeronautics and Space Administration
NASDAQ	National Association of Securities Dealers Automated Quotations
NATA	North American Telecommunications Association
NATO	North Atlantic Treaty Organization
NAU	Network accessible unit
NCP	Network control program
NCSA	National Center for Supercomputer Applications
NCTA	National Cable Television Association
NDIS	Network Driver Interface Specifications
NDSF	Non-dispersion-shifted fiber
NetBEUI	NetBIOS Extended User Interface
NetBIOS	Network Basic Input/Output System
NFS	Network File System (Sun)
NIC	Network Interface Card
NII	National Information Infrastructure
NIST	National Institute of Standards and Technology (formerly NBS)
NIU	Network interface unit
NLPID	Network Layer Protocol Identifier
NLSP	NetWare Link Services Protocol
NM	Network module
Nm	Nanometer
NMC	Network Management Center
NMS	Network Management System
NMT	Nordic Mobile Telephone
NMVT	Network Management Vector Transport Protocol
NNI	Network Node Interface; Network-to-Network Interface
NOC	Network Operations Center
NOCC	Network Operations Control Center

NOS	Network Operating System
NPA	Numbering Plan Area
NREN	National Research and Education Network
NRZ	Nonreturn to zero
NRZI	Nonreturn to zero inverted
NSA	National Security Agency
NSAP	Network service access point
NSAPA	Network service access point address
NSF	National Science Foundation
NTSC	National Television Systems Committee
NTT	Nippon Telephone and Telegraph
NVOD	Near Video on Demand
NZDSF	Non-zero dispersion-shifted fiber
OADM	Optical add-drop multiplexer
OAM	Operations, Administration, and Maintenance
OAM&P	Operations, Administration, Maintenance, and Provisioning
OAN	Optical area network
OBS	Optical burst switching
OC	Optical carrier
OEM	Original equipment manufacturer
O-E-O	Optical-electrical-optical
OLS	Optical Line System (Lucent)
OMAP	Operations, Maintenance, and Administration Part (SS7)
ONA	Open Network Architecture
ONU	Optical network unit
OOF	Out of frame
OS	Operating system
OSF	Open Software Foundation
OSI	Open Systems Interconnection (ISO, ITU-T)
OSI RM	Open Systems Interconnection Reference Model
OSPF	Open Shortest Path First (IETF)
OSS	Operation support systems
OTDM	Optical Time Division Multiplexing

OTDR	Optical Time-Domain Reflectometer
OUI	Organizationally Unique Identifier (SNAP)
OVD	Outside Vapor Deposition
OXC	Optical cross-connect
P/F	Poll/Final (HDLC)
PAD	Packet Assembler/Disassembler (X.25)
PAL	Phase alternate line
PAM	Pulse Amplitude Modulation
PANS	Pretty Amazing New Stuff
PBX	Private branch exchange
PCI	Pulse Code Modulation; Peripheral Component Interface
PCMCIA	Personal Computer Memory Card International Association
PCN	Personal Communications Network
PCS	Personal Communications Services
PDA	Personal digital assistant
PDH	Plesiochronous Digital Hierarchy
PDU	Protocol data unit
PIN	Positive-intrinsic-negative
PING	Packet Internet Groper (TCP/IP)
PLCP	Physical Layer Convergence Protocol
PLP	Packet Layer Protocol (X.25)
PM	Phase Modulation
PMD	Physical Medium Dependent (FDDI)
PNNI	Private Network Node Interface (ATM)
PON	Passive Optical Networking
POP	Point of presence
POSIT	Profiles for Open Systems Interworking Technologies
POSIX	Portable Operating System Interface for UNIX
POTS	Plain Old Telephone Service
PPP	Point-to-Point Protocol (IETF)
PRC	Primary reference clock
PRI	Primary rate interface
PROFS	Professional Office System

PROM	Programmable Read-Only Memory
PSDN	Packet Switched Data Network
PSK	Phase Shift Keying
PSPDN	Packet Switched Public Data Network
PSTN	Public Switched Telephone Network
PTI	Payload Type Identifier (ATM)
PTT	Post, Telephone, and Telegraph
PU	Physical unit (SNA)
PUC	Public Utility Commission
PVC	Permanent virtual circuit
QAM	Quadrature Amplitude Modulation
Q-bit	Qualified data bit (X.25)
QLLC	Qualified Logical Link Control (SNA)
QoS	Quality of service
QPSK	Quadrature Phase Shift Keying
QPSX	Queued Packet Synchronous Exchange
R&D	Research and Development
RADSL	Rate Adaptive Digital Subscriber Line
RAID	Redundant Array of Inexpensive Disks
RAM	Random Access Memory
RARP	Reverse Address Resolution Protocol (IETF)
RAS	Remote access server
RBOC	Regional Bell Operating Company
RF	Radio frequency
RFC	Request for Comments (IETF)
RFH	Remote Frame Handler (ISDN)
RFI	Radio Frequency Interference
RFP	Request for Proposal
RHC	Regional Holding Company
RHK	Ryan, Hankin, and Kent (Consultancy)
RIP	Routing Information Protocol (IETF)
RISC	Reduced Instruction Set Computer
RJE	Remote Job Entry

RNR	Receive Not Ready (HDLC)
ROM	Read-Only Memory
ROSE	Remote Operation Service Element
RPC	Remote Procedure Call
RPR	Resilient Packet Ring
RR	Receive Ready (HDLC)
RTS	Request to Send (EIA-232-E)
SAA	Systems Application Architecture (IBM)
SAAL	Signaling ATM Adaptation Layer (ATM)
SABM	Set Asynchronous Balanced Mode (HDLC)
SABME	Set Asynchronous Balanced Mode Extended (HDLC)
SAC	Single Attachment Concentrator (FDDI)
SAN	Storage Area Network
SAP	Service access point (generic)
SAPI	Service Access Point Identifier (LAPD)
SAR	Segmentation and Reassembly (ATM)
SAS	Single attachment station (FDDI)
SASE	Specific Applications Service Element (subset of CASE, Application Layer)
SATAN	System Administrator Tool for Analyzing Networks
SBS	Stimulated Brillouin scattering
SCCP	Signaling connection control point (SS7)
SCP	Service control point (SS7)
SCREAM™	Scalable Control of a Rearrangeable Extensible Array of Mirrors (Calient)
SCSI	Small Computer Systems Interface
SCTE	Serial Clock Transmit External (EIA-232-E)
SDH	Synchronous Digital Hierarchy (ITU-T)
SDLC	Synchronous Data Link Control (IBM)
S/DMS	SONET/Digital Multiplex System
SDS	Scientific Data Systems
SECAM	Sequential Color with Memory
SF	Superframe format (T-1)

SGML	Standard Generalized Markup Language
SGMP	Simple Gateway Management Protocol (IETF)
SHDSL	Symmetric HDSL
S-HTTP	Secure HTTP (IETF)
SIF	Signaling Information Field
SIG	Special interest group
SIO	Service Information Octet
SIP	Serial Interface Protocol
SIR	Sustained information rate (SMDS)
SLA	Service level agreement
SLIP	Serial Line Interface Protocol (IETF)
SM	Switching module
SMAP	System Management Application Part
SMDS	Switched Multimegabit Data Service
SMF	Single mode fiber
SMP	Simple Management Protocol; Switching module processor
SMR	Specialized Mobile Radio
SMS	Standard Management System (SS7)
SMTP	Simple Mail Transfer Protocol (IETF)
SNA	Systems Network Architecture (IBM)
SNAP	Subnetwork Access Protocol
S/N	Signal-to-Noise Ratio
SNI	Subscriber Network Interface (SMDS)
SNMP	Simple Network Management Protocol (IETF)
SNP	Sequence Number Protection
SOHO	Small-Office, Home-Office
SONET	Synchronous Optical Network
SPAG	Standards Promotion and Application Group
SPARC	Scalable Performance Architecture
SPE	Synchronous Payload Envelope (SONET)
SPID	Service Profile Identifier (ISDN)
SPM	Self-Phase Modulation
SPOC	Single point of contact

SPX	Sequenced Packet Exchange (NetWare)
SQL	Structured Query Language
SRB	Source Route Bridging
SRP	Spatial Reuse Protocol
SRS	Stimulated Raman scattering
SRT	Source Routing Transparent
SS7	Signaling System 7
SSL	Secure Socket Layer (IETF)
SSP	Service switching point (SS7)
SST	Spread Spectrum Transmission
STDM	Statistical Time Division Multiplexing
STM	Synchronous Transfer Mode; Synchronous Transport Module (SDH)
STP	Signal transfer point (SS7); Shielded twisted pair
STS	Synchronous Transport Signal (SONET)
STX	Start of text (BISYNC)
SVC	Signaling virtual channel (ATM); Switched virtual circuit
SXS	Step-by-step switching
SYN	Synchronization
SYNTRAN	Synchronous Transmission
TA	Terminal adapter (ISDN)
TAG	Technical Advisory Group
TASI	Time Assigned Speech Interpolation
TAXI	Transparent Asynchronous Transmitter/Receiver Interface (Physical Layer)
TCAP	Transaction Capabilities Application Part (SS7)
TCM	Time Compression Multiplexing; Trellis Coding Modulation
TCP	Transmission Control Protocol (IETF)
TDD	Time Division Duplexing
TDM	Time Division Multiplexing
TDMA	Time Division Multiple Access
TDR	Time Domain Reflectometer
TE1	Terminal Equipment type 1 (ISDN capable)

TE2	Terminal Equipment type 2 (non-ISDN capable)
TEI	Terminal Endpoint Identifier (LAPD)
TELRIC	Total Element Long-Run Incremental Cost
TIA	Telecommunications Industry Association
TIRKS	Trunk Integrated Record Keeping System
TL1	Transaction Language 1
TLAN	Transparent LAN
TM	Terminal multiplexer
TMN	Telecommunications Management Network
TMS	Time-multiplexed switch
TOH	Transport overhead (SONET)
TOP	Technical and Office Protocol
TOS	Type of service (IP)
TP	Twisted pair
TR	Token Ring
TRA	Traffic Routing Administration
TSI	Time Slot Interchange
TSLRIC	Total Service Long-Run Incremental Cost
TSO	Terminating screening office; Time-Sharing Option (IBM)
TSR	Terminate and Stay Resident
TSS	Telecommunication Standardization Sector (ITU-T)
TST	Time-Space-Time Switching
TSTS	Time-Space-Time-Space Switching
TTL	Time to live
TU	Tributary unit (SDH)
TUG	Tributary Unit Group (SDH)
TUP	Telephone User Part (SS7)
UA	Unnumbered acknowledgment (HDLC)
UART	Universal Asynchronous Receiver Transmitter
UBR	Unspecified bit rate (ATM)
UDI	Unrestricted Digital Information (ISDN)
UDP	User Datagram Protocol (IETF)
UHF	Ultra high frequency

UI	Unnumbered information (HDLC)
UNI	User-to-Network Interface (ATM, FR)
UNIT™	Unified Network Interface Technology™ (Ocular)
UNMA	Unified Network Management Architecture
UPS	Uninterruptable power supply
UPSR	Unidirectional Path-Switched Ring
UPT	Universal Personal Telecommunications
URL	Uniform Resource Locator
USART	Universal Synchronous Asynchronous Receiver Transmitter
USB	Universal serial bus
UTC	Coordinated universal time
UTP	Unshielded twisted pair (physical layer)
UUCP	UNIX-UNIX copy
VAN	Value-added network
VAX	Virtual Address Extension (DEC)
vBNS	Very High Speed Backbone Network Service
VBR	Variable bit rate (ATM)
VBR-NRT	Variable bit rate—Non-real-time (ATM)
VBR-RT	Variable bit rate—Real-time (ATM)
VC	Virtual channel (ATM); virtual circuit (PSN); virtual container (SDH)
VCC	Virtual channel connection (ATM)
VCI	Virtual channel identifier (ATM)
VCSEL	Vertical Cavity Surface Emitting Laser
VDSL	Very High-speed Digital Subscriber Line
VDSL	Very High bit rate Digital Subscriber Line
VERONICA	Very Easy Rodent-Oriented Netwide Index to Computerized Archives (Internet)
VGA	Variable graphics array
VHF	Very high frequency
VHS	Video home system
VID	VLAN ID
VINES	Virtual Networking System (Banyan)

VIP	VINES Internet Protocol
VLAN	Virtual LAN
VLF	Very low frequency
VLR	Visitor Location Register (Wireless/GSM)
VLSI	Very large scale integration
VM	Virtual machine (IBM); virtual memory
VMS	Virtual Memory System (DEC)
VOD	Video-on-demand
VP	Virtual path
VPC	Virtual path connection
VPI	Virtual Path identifier
VPN	Virtual private network
VR	Virtual reality
VSAT	Very small aperture terminal
VSB	Vestigial sideband
VSELP	Vector-Sum Excited Linear Prediction
VT	Virtual tributary
VTAM	Virtual Telecommunications Access Method (SNA)
VTOA	Voice and Telephony over ATM
VTP	Virtual Terminal Protocol (ISO)
WACK	Wait acknowledgment (BISYNC)
WACS	Wireless Access Communications System
WAIS	Wide Area Information Server (IETF)
WAN	Wide area network
WAP	Wireless Application Protocol (Wrong Approach to Portability)
WARC	World Administrative Radio Conference
WATS	Wide Area Telecommunications Service
WDM	Wavelength Division Multiplexing
WIN	Wireless In-building Network
WTO	World Trade Organization
WWW	World Wide Web (IETF)
WYSIWYG	What You See Is What You Get

xDSL	x-Type Digital Subscriber Line
XID	Exchange Identification (HDLC)
XNS	Xerox Network Systems
XPM	Cross-Phase Modulation
ZBTSI	Zero byte time slot interchange
ZCS	Zero code suppression

B

Glossary

3G   Third-generation (3G) systems provide access to a wide range of telecommunication services supported by both fixed telecommunication networks and other services specific to mobile users. A range of mobile terminal types will be supported and may be designed for mobile or fixed use. Key features of 3G systems are the compatibility of services, small terminals with worldwide roaming capabilities, Internet access and other multimedia applications, high bandwidth, and a wide range of services and terminals.

4G   Fourth-generation (4G) networks extend 3G network capacity by an order of magnitude, rely entirely on a packet infrastructure, use network elements that are 100 percent digital, and offer extremely high bandwidth.

Abend   A contraction of the words *abnormal end*, used to describe a computer crash in the mainframe world.

Absorption   A form of optical attenuation in which optical energy is converted into an alternative form, often heat. Often caused by impurities in the fiber, hydroxyl absorption is the best known form.

Acceptance angle   The critical angle within which incident light is totally internally reflected inside the core of an optical fiber.

Access   The set of technologies used to reach the network by a user.

Add-Drop Multiplexer (ADM)   A device used in SONET and SDH systems that has the capability to add and remove signal components without having to demultiplex the entire transmitted transmission stream, a significant advantage over legacy multiplexing systems such as DS3.

Aerial plant   Transmission equipment (including media, amplifiers, splice cases, and so on) that is suspended in the air between poles.

Alternate Mark Inversion   The encoding scheme used in T-1. Every other one is inverted in polarity from the one that preceded or follows it.

ALU   Arithmetic logic unit; the "brain" of a central processing unit (CPU) chip.

Amplifier   A device that increases the transmitted power of a signal. Amplifiers are typically spaced at carefully selected intervals along a transmission span.

Amplitude modulation   The process of causing an electromagnetic wave to carry information by changing or modulating the amplitude or loudness of the wave.

AMPS Advanced Mobile Phone Service; the modern analog cellular network.

Analog A signal that is continuously varying in time. Functionally, the opposite of digital.

Angular misalignment The reason for loss that occurs at the fiber ingress point. If the light source is improperly aligned with the fiber's core, some of the incident light will be lost, leading to a reduced signal strength.

Armor The rigid protective coating on some fiber cables that protects them from crushing and from chewing by rodents.

ASCII American Standards Code for Information Interchange. A 7-bit data encoding scheme.

ASIC Application-Specific Integrated Circuit, which is a specially designed IC created for a specific application.

Asynchronous Data that is transmitted between two devices that do not share a common clock source.

Asynchronous Transfer Mode (ATM) A standard for switching and multiplexing that relies on the transport of fixed-size data entities called cells, which are 53 octets in length. ATM has enjoyed a great deal of attention lately because its internal workings enable it to provide quality of service (QoS), a much-demanded option in modern data networks.

ATM Asynchronous Transfer Mode; one of the family of so-called fast packet technologies characterized by low error rates, high speed, and low cost. ATM is designed to connect seamlessly with SONET and SDH.

ATM Adaptation Layer (AAL) In ATM, the layer responsible for matching the payload being transported to a requested QoS level by assigning an ALL Type, which the network responds to.

Attenuation The reduction in signal strength in optical fiber that results from absorption and scattering effects.

Avalanche Photodiode (APD) An optical semiconductor receiver that has the capability to amplify weak, received optical signals by "multiplying" the number of received photons to intensify the strength of the received signal. APDs are used in transmission systems where receiver sensitivity is a critical issue.

Axis The center line of an optical fiber.

Back Scattering   The problem that occurs when light is scattered backward into the transmitter of an optical system. This impairment is analogous to echo, which occurs in copper-based systems.

Bandwidth   A measure of the number of bits per second that can be transmitted down a channel.

Baseband   In signaling, any technique that uses digital signal representation.

Baud   The *signaling rate* of a transmission system. This is one of the most misunderstood terms in all of telecommunications. Often used synonymously with bits per second, baud usually has a very different meaning. By using multibit encoding techniques, a single signal can simultaneously represent multiple bits. Thus, the bit rate can be many times the signaling rate.

Beam splitter   An optical device used to direct a single signal in multiple directions through the use of a partially reflective mirror or some form of an optical filter.

BECN   Backward Explicit Congestion Notification; a bit used in Frame Relay for notifying a device that it is transmitting too much information into the network and is therefore in violation of its service agreement with the switch.

Bell System Reference Frequency (BSRF)   In the early days of the Bell System, a single timing source in the Midwest provided a timing signal for all central office equipment in the country. This signal, delivered from a very expensive cesium clock source, was known as the BSRF. Today, GPS is used as the main reference clock source.

Bending loss   Loss that occurs when a fiber is bent far enough that its maximum allowable bend radius is exceeded. In this case, some of the light escapes from the waveguide, resulting in signal degradation.

Bend radius   The maximum degree to which a fiber can be bent before serious signal loss or fiber breakage occurs. Bend radius is one of the functional characteristics of most fiber products.

Bidirectional   A system that is capable of transmitting simultaneously in both directions.

Binary   A counting scheme that uses Base 2.

Bit rate   Bits per second.

Bluetooth   An open wireless standard designed to operate at a gross transmission level of 1 Mbps. Bluetooth is being positioned as a connectivity standard for personal area networks.

Bragg Grating   A device that relies on the formation of interference patterns to filter specific wavelengths of light from a transmitted signal. In optical systems, Bragg Gratings are usually created by wrapping a grating of the correct size around a piece of fiber that has been made photosensitive. The fiber is then exposed to strong ultraviolet light, which passes through the grating, forming areas of high and low refractive indices. Bragg Gratings (or filters, as they are often called) are used for selecting certain wavelengths of a transmitted signal, and are often used in optical switches, DWDM systems, and tunable lasers.

Broadband   Historically, broadband meant "any signal that is faster than the ISDN Primary Rate (T1 or E1). Today, it means "big pipe"—in other words, a very high transmission speed. In signaling, the term means analog.

Buffer   A coating that surrounds optical fiber in a cable and offers protection from water, abrasion, and so on.

Building Integrated Timing Supply (BITS)   The central office device that receives the clock signal from GPS or another source and feeds it to the devices in the office it controls.

Bus   The parallel cable that interconnects the components of a computer.

Butt splice   A technique in which two fibers are joined end to end by fusing them with heat or optical cement.

Cable   An assembly made up of multiple optical or electrical conductors as well as other inclusions such as strength members, waterproofing materials, armor, and so on.

Cable assembly   A complete optical cable that includes the fiber itself and terminators on each end to make it capable of attaching to a transmission or receive device.

Cable plant   The entire collection of transmission equipment in a system, including the signal emitters, the transport media, the switching and multiplexing equipment, and the receive devices.

Cable vault   The subterranean room in a central office where cables enter and leave the building.

Call center   A room in which operators receive calls from customers.

CCITT   Consultative Committee on International Telegraphy and Telephony. Now defunct and replaced by the ITU-TSS.

CDMA   Code Division Multiple Access, one of several digital cellular access schemes. CDMA relies on frequency hopping and noise modulation to encode conversations.

Cell   The standard protocol data unit in ATM networks. It comprises a 5-byte header and a 48-octet payload field.

Cell Loss Priority (CLP)   In ATM, a rudimentary single-bit field used to assign priority to transported payloads.

Cell Relay Service (CRS)   In ATM, the most primitive service offered by service providers, consisting of nothing more than raw bit transport with no assigned AAL types.

Cellular telephony   The wireless telephony system characterized by the following: low-power cells, frequency reuse, handoff, and central administration.

Center wavelength   The central operating wavelength of a laser used for data transmission.

Central office   A building that houses shared telephony equipment such as switches, multiplexers, and cable distribution hardware.

Central office terminal (COT)   In loop carrier systems, the device located in the central office that provides multiplexing and demultiplexing services. It is connected to the remote terminal.

Chained layers   The lower three layers of the OSI Model that provide for connectivity.

Chirp   A problem that occurs in laser diodes when the center wavelength shifts momentarily during the transmission of a single pulse. Chirp is due to the instability of the laser itself.

Chromatic dispersion   Because the wavelength of transmitted light determines its propagation speed in an optical fiber, different wavelengths of light will travel at different speeds during transmission. As a result, the multiwavelength pulse will tend to "spread out" during transmission, causing difficulties for the receive device. Material dispersion, waveguide dispersion, and profile dispersion all contribute to the problem.

CIR   Committed Information Rate; the volume of data that a Frame Relay provider absolutely guarantees it will transport for a customer.

Circuit Emulation Service (CES)   In ATM, a service that emulates private-line service by modifying (1) the number of cells transmitted per second and (2) the number of bytes of data contained in the payload of each cell.

Cladding   The fused silica "coating" that surrounds the core of an optical fiber. It typically has a different index of refraction than the core, causing light that escapes from the core into the cladding to be refracted back into the core.

CLEC Competitive local exchange carrier; a small telephone company that competes with the incumbent player in its own marketplace.

CMOS Complimentary Metal Oxide Semiconductor, a form of integrated circuit technology that is typically used in low-speed and low-power applications.

Coating The plastic substance that covers the cladding of an optical fiber. It is used to prevent damage to the fiber itself through abrasion.

Coherent A form of emitted light in which all the rays of the transmitted light align themselves with the same transmission axis, resulting in a narrow, tightly focused beam. Lasers emit coherent light.

Compression The process of reducing the size of a transmitted file without losing the integrity of the content by eliminating redundant information prior to transmitting or storing.

Concatenation The technique used in SONET and SDH in which multiple payloads are "ganged" together to form a super-rate frame capable of transporting payloads greater in size than the basic transmission speed of the system. Thus, an OC-12c provides 622.08 Mbps of total bandwidth, as opposed to an OC-12, which also offers 622.08 Mbps, but in increments of OC-1 (51.84 Mbps).

Conditioning The process of doctoring a dedicated circuit to eliminate the known and predictable results of distortion.

Congestion The condition that results when traffic arrives faster than it can be processed by a server.

Connectivity The process of providing the electrical transport of data.

Connector A device, usually mechanical, used to connect a fiber to a transmit or receive device or to bond two fibers.

Core The central high-speed transport region of the network that provides the primary transmission path for an optical signal. It usually has a higher index of refraction than the cladding.

Counter-rotating ring A form of transmission system that comprises two rings operating in opposite directions. Typically, one ring serves as the active path while the other serves as the protect or backup path.

CPU Central processing unit. Literally, the chipset in a computer that provides the intelligence.

CRC   Cyclic redundancy check; a mathematical technique for checking the integrity of the bits in a transmitted file.

Critical angle   The angle at which total internal reflection occurs.

Cross-Phase Modulation (XPM)   A problem that occurs in optical fiber that results from the nonlinear index of refraction of the silica in the fiber. Because the index of refraction varies according to the strength of the transmitted signal, some signals interact with each other in destructive ways. XPM is considered to be a fiber nonlinearity.

CSMA/CD   Carrier Sense Multiple Access with Collision Detection; the medium access scheme used in Ethernet LANs and characterized by an "if it feels good, do it" approach.

Customer Relationship Management (CRM)   A technique for managing the relationship between a service provider and a customer through the discrete management of knowledge about the customer.

Cutoff wavelength   The wavelength below which single mode fiber ceases to be single mode.

Cylinder   A stack of tracks to which data can be logically written on a hard drive.

Dark fiber   Optical fiber that is sometimes leased to a client that is not connected to a transmitter or receiver. In a dark fiber installation, it is the customer's responsibility to terminate the fiber.

Data   Raw, unprocessed zeroes and ones.

Data communications   The science of moving data between two or more communicating devices.

Datagram   The service provided by a connectionless network. Often said to be unreliable, this service makes no guarantees with regard to latency or sequentiality.

Data mining   A technique in which enterprises extract information about customer behavior by analyzing data contained in their stored transaction records.

DCE   Data Circuit Terminating Equipment; a modem or other device that delineates the end of the service provider's circuit.

DE   Discard Eligibility bit; a primitive single-bit technique for prioritizing traffic that is to be transmitted.

Decibel (dB)   A logarithmic measure of the strength of a transmitted signal. Because it is a logarithmic measure, a 20 dB loss would indicate that the received signal is one one-hundredth its original strength.

Dense Wavelength Division Multiplexing (DWDM) A form of frequency division multiplexing (FDM) in which multiple wavelengths of light are transmitted across the same optical fiber. These DWDM systems typically operate in the so-called L-Band (1625 nm) and have channels that are spaced between 50 and 100 GHz apart. Newly announced products may dramatically reduce this spacing.

Detector An optical receive device that converts an optical signal into an electrical signal so that it can be handed off to a switch, router, multiplexer, or other electrical transmission device. These devices are usually either NPN or APDs.

Diameter mismatch loss Loss that occurs when the diameter of a light emitter and the diameter of the ingress fiber's core are dramatically different.

Dichroic filter A filter that transmits light in a wavelength-specific fashion, reflecting nonselected wavelengths.

Dielectric A substance that is nonconducting.

Diffraction grating A grid of closely spaced lines that are used to selectively direct specific wavelengths of light as required.

Digital A signal characterized by discrete states. The opposite of analog.

Digital hierarchy In North America, the multiplexing hierarchy that enables 64 Kbps DS-0 signals to be combined to form DS-3 signals for a high bit rate transport.

Digital Subscriber Line Access Multiplexer (DSLAM) The multiplexer in the central office that receives voice and data signals on separate channels, relaying voice to the local switch and data to a router elsewhere in the office.

Diode A semiconductor device that only enables a current to flow in a single direction.

Direct attached storage (DAS) A storage option in which the storage media (hard drives, CDs, and so on) are either integral to the server (internally mounted) or are directly connected to one of the servers.

Dispersion The spreading of a light signal over time that results from modal or chromatic inefficiencies in the fiber.

Dispersion compensating fiber (DCF) A segment of fiber that exhibits the opposite dispersion effect of the fiber to which it is coupled. DCF is used to counteract the dispersion of the other fiber.

Dispersion-Shifted Fiber (DSF) A form of optical fiber that is designed to exhibit zero dispersion within the C-Band (1550 nm). DSF does not work well for DWDM because of Four Wave Mixing problems; Non-Zero Dispersion Shifted Fiber is used instead.

Distortion   A known and measurable (and therefore correctable) impairment on transmission facilities.

Dopant   Substances used to lower the refractive index of the silica used in optical fiber.

DS-0   Digital signal level 0, a 64 Kbps signal.

DS-1   Digital signal level 1, a 1.544 Mbps signal.

DS-2   Digital signal level 2, a 6.312 Mbps signal.

DS-3   A 44.736 Mbps signal format found in the North American Digital Hierarchy.

DSL   Digital Subscriber Line, a technique for transporting high-speed digital data across the analog local loop while (in some cases) transporting voice simultaneously.

DTE   Data Terminal Equipment; user equipment that is connected to a DCE.

DTMF   Dual-Tone, Multi-Frequency; the set of tones used in modern phones to signal dialed digits to the switch. Each button triggers a pair of tones.

Duopoly   The current regulatory model for cellular systems; two providers are assigned to each market. One is the wireline provider (typically the local ILEC), the other an independent provider.

DWDM   Dense Wavelength Division Multiplexing; a form of frequency division multiplexing that enables multiple optical signals to be transported simultaneously across a single fiber.

E1   The 2.048 Mbps transmission standard found in Europe and other parts of the world. It is analogous to the North American T1.

EBCDIC   Extended Binary Coded Decimal Interchange Code; an 8-bit data encoding scheme.

Edge   The periphery of the network where aggregation, QoS, and IP implementation take place. This is also where most of the intelligence in the network resides.

EDGE   Enhanced Data for Global Evolution; a 384 Kbps enhancement to GSM.

Edge-emitting diode   A diode that emits light from the edge of the device rather than the surface, resulting in a more coherent and directed beam of light.

Effective area   The cross-section of a single-mode fiber that carries the optical signal.

EIR   Excess information rate; the amount of data that is being transmitted by a user above the CIR in Frame Relay.

Encryption   The process of modifying a text or image file to prevent unauthorized users from viewing the content.

End-to-end layers   The upper four layers of the OSI Model that provide interoperability.

Enterprise Resource Planning (ERP)   A technique for managing customer interactions through data mining, knowledge management, and customer relationship management (CRM).

Erbium-Doped Fiber Amplifier (EDFA)   A form of optical amplifier that uses the element erbium to bring about the amplification process. Erbium has the enviable quality that when struck by light operating at 980 nm, it emits photons in the 1550 nm range, thus providing agnostic amplification for signals operating in the same transmission window.

ESF   Extended superframe; the framing technique used in modern T-carrier systems that provides a dedicated data channel for the nonintrusive testing of customer facilities.

Ethernet   A LAN product developed by Xerox that relies on a CSMA/CD medium access scheme.

Evanescent wave   Light that travels down the inner layer of the cladding instead of down the fiber core.

Extrinsic loss   Loss that occurs at splice points in an optical fiber.

Eye pattern   A measure of the degree to which bit errors are occurring in optical transmission systems. The width of the "eyes" (eye patterns look like figure eights lying on their sides) indicates the relative bit error rate.

Facility   A circuit.

Faraday Effect   Sometimes called the magneto-optical effect, the Faraday Effect describes the degree to which some materials can cause the polarization angle of incident light to change when placed within a magnetic field that is parallel to the propagation direction.

Fast Ethernet   A version of Ethernet that operates at 100 Mbps.

Fast packet   Technologies characterized by low error rates, high speed, and low cost.

FDMA   Frequency Division Multiple Access; the access technique used in analog AMPS cellular systems.

FEC Forward Error Correction; an error correction technique that sends enough additional overhead information along with the transmitted data so that a receiver not only detects an error, but actually fixes it without requesting a resend.

FECN Forward Explicit Congestion Notification; a bit in the header of a Frame Relay frame that can be used to notify a distant switch that the frame experienced severe congestion on its way to the destination.

Ferrule A rigid or semirigid tube that surrounds optical fibers and protects them.

Fiber grating A segment of photosensitive optical fiber that has been treated with ultraviolet light to create a refractive index within the fiber that varies periodically along its length. It operates analogously to a fiber grating and is used to select specific wavelengths of light for transmission.

Fiber-to-the-Curb (FTTC) A transmission architecture for service delivery in which a fiber is installed in a neighborhood and terminated at a junction box. From there, coaxial cable or twisted pair can be cross-connected from the O-E converter to the customer premises. If coax is used, the system is called Hybrid Fiber Coax (HFC); twisted pair-based systems are called Switched Digital Video (SDV).

Fiber-to-the-Home (FTTH) Similar to FTTC, except that FTTH extends the optical fiber all the way to the customer premises.

Fibre Channel A set of standards for a serial I/O bus that supports a range of port speeds, including 133 Mbps, 266 Mbps, 530 Mbps, 1 Gbps, and, soon, 4 Gbps. The standard supports point-to-point connections, switched topologies, and an arbitrated loop architecture.

Four Wave Mixing (FWM) The nastiest of the so-called fiber nonlinearities. FWM is commonly seen in DWDM systems and occurs when the closely spaced channels mix and generate the equivalent of optical sidebands. The number of these sidebands can be expressed by the equation $1/2(n^3 - n^2)$, where n is the number of original channels in the system. Thus, a 16-channel DWDM system will potentially generate 1,920 interfering sidebands!

Frame A variable size data transport entity.

Frame Relay One of the family of so-called fast packet technologies characterized by low error rates, high speed, and low cost.

Frame Relay Bearer Service (FRBS) In ATM, a service that enables Frame Relay frame to be transported across an ATM network.

Freespace optics   A metro transport technique that uses a narrow, unlicensed optical beam to transport high-speed data.

Frequency-Agile   The capability of a receiving or transmitting device to change its frequency in order to take advantage of alternate channels.

Frequency-Division Multiplexing   The process of assigning specific frequencies to specific users.

Frequency modulation   The process of causing an electromagnetic wave to carry information by changing or modulating the frequency of the wave.

Fresnel loss   The loss that occurs at the interface between the head of the fiber and the light source to which it is attached. At air-glass interfaces, the loss usually equates to about 4 percent.

Full-duplex   Two-way simultaneous transmission.

Fused fiber   A group of fibers that are fused together so that they will remain in alignment. They are often used in one-to-many distribution systems for the propagation of a single signal to multiple destinations. Fused fiber devices play a key role in passive optical networking (PON).

Fusion splice   A splice made by melting the ends of the fibers together.

Generic Flow Control (GFC)   In ATM, the first field in the cell header. It is largely unused except when it is overwritten in NNI cells, in which case it becomes additional space for virtual path addressing.

GEOS   Geosynchronous Earth Orbit Satellite; a family of satellites that orbit above the equator at an altitude of 22,300 miles and provide data and voice transport services.

Gigabit Ethernet   A version of Ethernet that operates at 1,000 Mbps.

Global Positioning System (GPS)   The array of satellites used for radio locations around the world. In the telephony world, GPS satellites provide an accurate timing signal for synchronizing office equipment.

Go-Back-N   A technique for error correction that causes all frames of data to be transmitted again, starting with the errored frame.

Gozinta   "Goes into."

Gozouta   "Goes out of."

GPRS   General Packet Radio Service; another add-on for GSM networks that is not enjoying a great deal of success in the market yet. Stay tuned.

Graded Index Fiber (GRIN)   A type of fiber in which the refractive index changes gradually between the central axis of the fiber and the outer layer, instead of abruptly at the core-cladding interface.

Groom (and Fill)   Similar to add-drop, groom and fill refers to the ability to add (fill) and drop (groom) payload components at intermediate locations along a network path.

GSM   Global System for Mobile Communications; the wireless access standard used in many parts of the world that offers two-way paging, short messaging, and two-way radio in addition to cellular telephony.

GUI   Graphical User Interface; the computer interface characterized by the "click, move, drop" method of file management.

Half-duplex   Two-way transmission, but only one direction at a time.

Haptics   The science of providing tactile feedback to a user electronically. Often used in high-end virtual reality systems.

Headend   The signal origination point in a cable system.

Header   In ATM, the first five bytes of the cell. The header contains information used by the network to route the cell to its ultimate destination. Fields in the cell header include Generic Flow Control, Virtual Path Identifier, Virtual Channel Identifier, Payload Type Identifier, Cell Loss Priority, and Header Error Correction.

Header Error Correction (HEC)   In ATM, the header field used to recover from bit errors in the header data.

Hertz (Hz)   A measure of cycles per second in transmission systems.

Hop count   A measure of the number of machines a message or packet has to pass through between the source and the destination. Often used as a diagnostic tool.

Hybrid Fiber Coax   A transmission system architecture in which a fiber feeder penetrates a service area and is then cross-connected to coaxial cable feeders into the customers' premises.

ILEC   Incumbent local exchange carrier; an RBOC.

Index of refraction   A measure of the ratio between the velocity of light in a vacuum and the velocity of the same light in an optical fiber. The refractive index is always greater than one and is denoted n.

Information   Data that has been converted to a manipulable form.

Infrared (IR)   The region of the spectrum within which most optical transmission systems operate, found between 700 nm and 0.1 mm.

Injection laser   A semiconductor laser (synonym).

Inside plant   Telephony equipment that is outside of the central office.

Intermodulation   A fiber nonlinearity that is similar to four-wave mixing, in which the power-dependent refractive index of the transmission medium enables signals to mix and create destructive sidebands.

Interoperability   The ability to logically share information between two communicating devices and be able to read and understand the data of the other. In SONET and SDH, this is the ability of devices from different manufacturers to send and receive information to and from each other successfully.

Intrinsic loss   A loss that occurs as the result of physical differences in the two fibers being spliced.

ISDN   Integrated Services Digital Network; a digital local loop technology that offers moderately high bit rates to customers.

Isochronous   A situation in a timing system when a constant delay exists across a network.

ISP   Internet Service Provider; a company that offers Internet access.

ITU   International Telecommunications Union; a division of the United Nations that is responsible for managing the telecomm standards development and maintenance processes.

ITU-TSS   ITU Telecommunications Standardization Sector; the ITU organization responsible for telecommunications standards development.

Jacket   The protective outer coating of an optical fiber cable. The jacket may be polyethylene, Kevlar®, or metallic.

JPEG   Joint Photographic Experts Group; a standards body tasked with developing standards for the compression of still images.

Jumper   An optical cable assembly, usually fairly short, that is terminated on both ends with connectors.

Knowledge   Information that has been acted upon and modified through some form of intuitive human thought process.

Knowledge management   The process of managing all that a company knows about its customers in an intelligent way so that some benefit is attained for both the customer and the service provider.

Lambda   A single wavelength on a multichannel DWDM system.

LAN   Local area network; a small network that has the following characteristics: privately owned, high speed, low error rate, and physically small.

LAN Emulation (LANE)   In ATM, a service that defines the capability to provide bridging services between LANs across an ATM network.

Large-core fiber   Fiber that characteristically has a core diameter of 200 microns or more.

Laser   An acronym for Light Amplification by the Stimulated Emission of Radiation. Lasers are used in optical transmission systems because they produce coherent light that is almost purely monochromatic.

Laser diode (LD)   A diode that produces coherent light when a forward biasing current is applied to it.

LATA   Local Access and Transport Area; the geographic area within which an ILEC is allowed to transport traffic. Beyond LATA boundaries, the ILEC must hand traffic off to a long-distance carrier.

LEOS   Low earth orbit satellite; satellites that orbit pole to pole instead of above the equator and offer near-instantaneous response time.

Light-emitting diode (LED)   A diode that emits incoherent light when a forward bias current is applied to it. LEDs are typically used in shorter-distance, lower-speed systems.

Lightguide   A term that is used synonymously with optical fiber.

Line Overhead (LOH)   In SONET, the overhead that is used to manage the network regions between multiplexers.

Linewidth   The spectrum of wavelengths that make up an optical signal.

Load coil   A device that tunes the local loop to the voiceband.

Local loop   The pair of wires (or digital channel) that runs between the customer's phone (or computer) and the switch in the local central office.

Loose tube optical cable   An optical cable assembly in which the fibers within the cable are loosely contained within tubes inside the sheath of the cable. The fibers are able to move within the tube, thus allowing them to adapt and move without damage as the cable is flexed and stretched.

Loss   The reduction in signal strength that occurs over distance, usually expressed in decibels.

M13   A multiplexer that interfaces between DS-1 and DS-3 systems.

Mainframe   A large computer that offers support for very large databases and large numbers of simultaneous sessions.

MAN   Metropolitan area network; a network, larger than a LAN, that provides high-speed services within a metropolitan area.

Material dispersion A dispersion effect caused by the fact that different wavelengths of light travel at different speeds through a medium.

MDF Main Distribution Frame; the large iron structure that provides physical support for cable pairs in a central office between the switch and the incoming/outgoing cables.

Message switching An older technique that sends entire messages from point to point instead of breaking the message into packets.

Metasignaling virtual channel (MSVC) In ATM, a signaling channel that is always on. It is used for the establishment of temporary signaling channels as well as channels for voice and data transport.

Metropolitan Optical Network (MON) An all-optical network deployed in a metro region.

Microbend Changes in the physical structure of an optical fiber caused by bending that can result in light leakage from the fiber.

Midspan meet In SONET and SDH, the term used to describe interoperability. *See also* interoperability.

Modal dispersion *See* multimode dispersion.

Mode A single wave that propagates down a fiber. Multimode fiber enables multiple modes to travel, while single-mode fiber enables only a single mode to be transmitted.

Modem A term from the words modulate and demodulate. Its job is to make a computer appear to the network like a telephone.

Modulation The process of changing or modulating a carrier wave to cause it to carry information.

MPEG Moving Picture Experts Group; a standards body tasked with crafting standards for motion pictures.

MPLS A level three protocol designed to provide QoS across IP networks without the need for ATM by assigning QoS labels to packets as they enter the network.

MTSO Mobile Telephone Switching Office; a central office with special responsibilities for handling cellular services and the interface between cellular users and the wireline network.

Multidwelling Unit (MDU) A building that houses multiple residence customers such as an apartment building.

Multimode dispersion Sometimes referred to as modal dispersion, multimode dispersion is caused by the fact that different modes take different times to move from the ingress point to the egress point of a fiber, thus resulting in modal spreading.

Multimode fiber   Fiber that has a core diameter of 62.5 microns or greater, wide enough to enable multiple modes of light to be simultaneously transmitted down the fiber.

Multiplexer   A device that has the capability to combine multiple inputs into a single output as a way to reduce the requirement for additional transmission facilities.

Multiprotocol over ATM (MPOA)   In ATM, a service that allows IP packets to be routed across an ATM network.

Multitenant unit (MTU)   A building that houses multiple enterprise customers such as a high-rise office building.

Near-End Crosstalk (NEXT)   The problem that occurs when an optical signal is reflected back toward the input port from one or more output ports. This problem is sometimes referred to as isolation directivity.

Network-attached storage (NAS)   An architecture in which a server accesses storage media via a LAN connection. The storage media are connected to another server.

Noise   An unpredictable impairment in networks. It cannot be anticipated; it can only be corrected after the fact.

Non-Dispersion Shifted Fiber (NDSF)   Fiber that is designed to operate at the low-dispersion, second operational window (1310 nm).

Non-Zero Dispersion-Shifted Fiber (NZDSF)   A form of single-mode fiber that is designed to operate just outside the 1550 nm window so that fiber nonlinearities, particularly FWM, are minimized.

Numerical aperture (NA)   A measure of the capability of a fiber to gather light, NA is also a measure of the maximum angle at which a light source can be from the center axis of a fiber in order to collect light.

OAM&P   Operations, Administration, Maintenance and Provisioning; the four key areas in modern network management systems. OAM&P was first coined by the Bell System and continues in widespread use today.

OC-n   Optical Carrier level n, a measure of bandwidth used in SONET systems. OC-1 is 51.84 Mbps; OC-n is n times 51.84 Mbps.

Optical amplifier   A device that amplifies an optical signal without first converting it to an electrical signal.

Optical Burst Switching (OBS)   A technique that uses a "one-way reservation" technique so that a burst of user data, such as a cluster of IP packets, can be sent without having to establish a dedicated path prior to transmission. A control packet is sent first to reserve the wavelength, followed by the traffic burst. As a result, OBS avoids the protracted end-to-end setup delay and also improves the utilization of optical channels for variable bit rate services.

Optical Carrier Level n (OC-n)   In SONET, the transmission level at which an optical system is operating.

Optical isolator   A device used to selectively block specific wavelengths of light.

Optical Time Domain Reflectometer (OTDR)   A device used to detect failures in an optical span by measuring the amount of light reflected back from the air-glass interface at the failure point.

OSS   Operations Support Systems; another term for OAM&P.

Outside Plant   Telephone equipment that is outside the central office.

Overhead   That part of a transmission stream that the network uses to manage and direct the payload to its destination.

Packet   A variable size entity normally carried inside a frame or cell.

Packet switching   The technique for transmitting packets across a WAN.

Path overhead   In SONET and SDH, that part of the overhead that is specific to the payload being transported.

Payload   In SONET and SDH, the user data that is being transported.

Payload Type identifier (PTI)   In ATM, a cell header field that is used to identify network congestion and cell type. The first bit indicates whether the cell was generated by the user or by the network, while the second indicates the presence or absence of congestion in user-generated cells, or flow-related OAM information in cells generated by the network. The third bit is used for service-specific, higher-layer functions in the user-to-network direction, such as to indicate that a cell is the last in a *series* of cells. From the network to the user, the third bit is used with the second bit to indicate whether the OAM information refers to segment or end-to-end-related information flow.

PBX   Private branch exchange; literally a small telephone switch located on the customer premises. The PBX connects back to the service provider's central office via a collection of high-speed trunks.

PCM   Pulse code modulation; the encoding scheme used in North America for digitizing voice.

Phase modulation   The process of causing an electromagnetic wave to carry information by changing or modulating the phase of the wave.

Photodetector   A device used to detect an incoming optical signal and convert it to an electrical output.

Photodiode   A semiconductor that converts light to electricity.

Photon   The fundamental unit of light, sometimes referred to as a quantum of electromagnetic energy.

Photonic   The optical equivalent of the term electronic.

Pipelining   The process of having multiple, unacknowledged, outstanding messages in a circuit between two communicating devices.

Pixel   Contraction of the terms picture element. The tiny color elements that make up the screen on a computer monitor.

Planar waveguide   A waveguide fabricated from a flat material such as a sheet of glass, into which are etched fine lines used to conduct optical signals.

Plenum   The air-handling space in buildings found inside walls, under floors, and above ceilings. The plenum spaces are often used as conduits for optical cables.

Plenum cable   Cable that passes fire-retardant tests so that it can legally be used in plenum installations.

Plesiochronous   In timing systems, a term that means "almost synchronized." It refers to the fact that in SONET and SDH systems, payload components frequently derive from different sources and therefore may have slightly different phase characteristics.

Pointer   In SONET and SDH, a field that is used to indicate the beginning of the transported payload.

Polarization   The process of modifying the direction of the magnetic field within a light wave.

Polarization Mode Dispersion (PMD)   The problem that occurs when light waves with different polarization planes in the same fiber travel at different velocities down the fiber.

Preform   The cylindrical mass of highly pure fused silica from which optical fiber is drawn during the manufacturing process. In the industry, the preform is sometimes referred to as a gob.

Private line   A dedicated point-to-point circuit.

Protocol   A set of rules that facilitates communications.

Pulse spreading   The widening or spreading out of an optical signal that occurs over distance in a fiber.

Pump laser   The laser that provides the energy used to excite the dopant in an optical amplifier.

PVC   Permanent virtual circuit; a circuit provisioned in Frame Relay or ATM that does not change without service order activity by the service provider.

Q.931   The set of standards that defines signaling packets in ISDN networks.

Quantize   The process of assigning numerical values to the digitized samples created as part of the voice digitization process.

RAM   Random Access Memory; the volatile memory used in computers for short-term storage.

Rayleigh scattering   A scattering effect that occurs in optical fiber as the result of fluctuations in silica density or chemical composition. Metal ions in the fiber often cause Rayleigh scattering.

RBOC   Regional Bell Operating Company; today called an ILEC.

Refraction   The change in direction that occurs in a light wave as it passes from one medium into another. The most common example is the bending that is often seen to occur when a stick is inserted into water.

Refractive index   A measure of the speed at which light travels through a medium, usually expressed as a ratio compared to the speed of the same light in a vacuum.

Regenerative repeater   A device that reconstructs and regenerates a transmitted signal that has been weakened over distance.

Regenerator   A device that recreates a degraded digital signal before transmitting it on to its final destination.

Remote terminal (RT)   In loop carrier systems, the multiplexer located in the field. It communicates with the central office terminal (COT).

Repeater   *See* regenerator.

Resilient Packet Ring (RPR)   A ring architecture that comprises multiple nodes that share access to a bidirectional ring. Nodes send data across the ring using a specific MAC protocol created for RPR. The goal of the RPR topology is to interconnect multiple nodes' ring architecture that is media independent for efficiency purposes.

ROM   Read-Only Memory that cannot be erased; it is often used to store critical files or boot instructions.

Scattering   The "backsplash" or reflection of an optical signal that occurs when it is reflected by small inclusions or particles in the fiber.

SDH   The abbreviation for Synchronous Digital Hierarchy, the European equivalent of SONET.

Section overhead (SOH)   In SONET systems, the overhead that is used to manage the network regions that occur between repeaters.

Sector   A quadrant on a disk drive to which data can be written. Used for locating information on the drive.

Selective retransmit   An error correction technique in which only the errored frames are retransmitted.

Self-Phase Modulation (SPM)   A modulation process in which the refractive index of glass is directly related to the power of the transmitted signal. As the power fluctuates, so too does the index of refraction, causing waveform distortion.

Sheath   One of the layers of protective coating in an optical fiber cable.

Signaling   The techniques used to set up, maintain, and tear down a call.

Signaling virtual channel (SVC)   In ATM, a temporary signaling channel used to establish paths for the transport of user traffic.

Simplex   One way transmission *only*.

Single-Mode Fiber (SMF)   The most popular form of fiber today, characterized by the fact that it enables only a single mode of light to propagate down the fiber.

Soliton   A unique waveform that takes advantage of nonlinearities in the fiber medium, the result of which is a signal that suffers essentially no dispersion effects over long distances. Soliton transmission is an area of significant study at the moment, because of the promise it holds for long-haul transmission systems.

SONET   Abbreviation for the Synchronous Optical Network, a multiplexing standard that begins at DS3 and provides standards-based multiplexing up to gigabit speeds. SONET is widely used in telephone company long-haul transmission systems and was one of the first widely deployed optical transmission systems.

Source   The emitter of light in an optical transmission system.

Spatial Reuse Protocol (SRP)   A ring-based protocol that allows routers to be interconnected by dual fiber rings in a highly survivable configuration.

SS7   Signaling System Seven, the current standard for telephony signaling worldwide.

Standards   The published rules that govern an industry's activities.

Step index fiber   Fiber that exhibits a continuous refractive index in the core and that "steps" at the core-cladding interface.

Stimulated Brillouin scattering (SBS)   A fiber nonlinearity that occurs when a light signal traveling down a fiber interacts with acoustic vibrations in the glass matrix (sometimes called photon-phonon interaction), causing light to be scattered or reflected back toward the source.

Stimulated Raman scattering (SRS)   A fiber nonlinearity that occurs when power from short-wavelength, high-power channels is bled into longer-wavelength, lower-power channels.

Storage Area Network (SAN)   A dedicated storage network that provides access to stored content. In a SAN, multiple servers may have access to the same servers.

Store and forward   The transmission technique in which data is transmitted to a switch, stored there, examined for errors, examined for address information, and forwarded on to the final destination.

Strength member   The strand within an optical cable that is used to provide tensile strength to the overall assembly. The member is usually composed of steel, fiberglass, or Aramid yarn.

Surface-emitting diode   A semiconductor that emits light from its surface, resulting in a low-power, broad-spectrum emission.

SVC   A Frame Relay or ATM technique in which a customer can establish on-demand circuits as required.

Synchronous   A term that means that both communicating devices derive their synchronization signal from the same source.

Synchronous Transmission Signal Level 1 (STS-1)   In SONET systems, the lowest transmission level in the hierarchy. STS is the electrical equivalent of OC.

T1   The 1.544 Mbps transmission standard in North America.

T3   In the North American Digital Hierarchy, a 44.736 Mbps signal.

Tandem   A switch that serves as an interface between other switches and typically does not directly host customers.

TDMA   Time Division Multiple Access; a digital technique for cellular access in which customers share access to a frequency on a round-robin, time division basis.

Telecommunications   The science of transmitting sound over distance.

Terminal multiplexer   In SONET and SDH systems, a device that is used to distribute payload to or receive payload from user devices at the end of an optical span.

Tight buffer cable An optical cable in which the fibers are tightly bound by the surrounding material.

Time-Division Multiplexing The process of assigning timeslots to specific users.

Token Ring A LAN technique, originally developed by IBM, that uses token passing to control access to the shared infrastructure.

Total internal reflection The phenomenon that occurs when light strikes a surface at such an angle that all the light is reflected back into the transporting medium. In optical fiber, total internal reflection is achieved by keeping the light source and the fiber core oriented along the same axis so that the light that enters the core is reflected back into the core at the core-cladding interface.

Transceiver A device that incorporates both a transmitter and a receiver in the same housing, thus reducing the need for rack space.

Transponder A device that incorporates a transmitter, a receiver, and a multiplexer on a single chassis.

Twisted pair The wire used to interconnect customers to the telephone network.

UPS Uninterruptible Power Supply; part of the central office power plant that prevents power outages.

Vertical Cavity Surface Emitting Laser (VCSEL) A small, highly efficient laser that emits light vertically from the surface of the wafer on which it is made.

Virtual channel (VC) In ATM, a unidirectional channel between two communicating devices.

Virtual channel identifier (VCI) In ATM, the field that identifies a virtual channel.

Virtual container In SDH, the technique used to transport subrate payloads.

Virtual path (VP) In ATM, a combination of unidirectional virtual channels that make up a bidirectional channel.

Virtual path identifier (VPI) In ATM, the field that identifies a virtual path.

Virtual private network A network connection that provides private-like services over a public network.

Virtual tributary (VT) In SONET, the technique used to transport subrate payloads.

Voiceband   The 300–3300 Hz band used for the transmission of voice traffic.

Voice/Telephony over ATM (VTOA)   In ATM, a service used to transport telephony signals across an ATM network.

WAN   Wide area network; a network that provides connectivity over a large geographical area.

Waveguide   A medium that is designed to conduct light within itself over a significant distance, such as optical fiber.

Waveguide dispersion   A form of chromatic dispersion that occurs when some of the light traveling in the core escapes into the cladding, traveling there at a different speed than the light in the core.

Wavelength   The distance between the same points on two consecutive waves in a chain, such as from the peak of wave one to the peak of wave two. Wavelength is related to frequency by the equation where lambda (λ) is the wavelength, c is the speed of light, and f is the frequency of the transmitted signal.

Wavelength Division Multiplexing (WDM)   The process of transmitting multiple wavelengths of light down a fiber.

Window   A region within which optical signals are transmitted at specific wavelengths to take advantage of propagation characteristics that occur there, such as minimum loss or dispersion.

Window size   A measure of the number of messages that can be outstanding at any time between two communicating entities.

Zero Dispersion Wavelength   The wavelength at which material and waveguide dispersion cancel each other.

INDEX

B

M

ABOUT THE AUTHOR

Steven Shepard is a professional writer and educator specializing in international telecommunications. He is the author of five well-received books: *Telecommunications Convergence*; *SONET/SDH Demystified*; *Optical Networking Crash Course*; *Telecom Crash Course*; and the recently published *Video Devices Demystified*. Formerly with Hill Associates and now president of the Shepard Communications Group, he conducts seminars and workshops on telecom topics around the country. He lives in Williston, Vermont.